PHYSICAL GEOGRAPHY Made Simple

The Made Simple series
has been created
primarily for self-education
but can equally well
be used as
an aid to group study.
However complex the subject,
the reader is taken
step by step,
clearly and methodically,
through the course. Each volume
has been prepared by experts,
using throughout the
Made Simple technique of teaching.
Consequently the gaining
of knowledge now becomes
an experience to be enjoyed.

Accounting	Electronics
Acting and Stagecraft	English
Additional Mathematics	French
Advertising	Geology
Anthropology	German
Applied Economics	Human Anatomy
Applied Mathematics	Italian
Applied Mechanics	Journalism
Art Appreciation	Latin
Art of Speaking	Law
Art of Writing	Management
Biology	Marketing
Book-keeping	Mathematics
British Constitution	Modern European History
Calculus	New Mathematics
Chemistry	Office Practice
Childcare	Organic Chemistry
Commerce	Philosophy
Commercial Law	Photography
Company Administration	Physical Geography
Computer Programming	Physics
Cookery	Pottery
Cost and Management Accounting	Psychology
Data Processing	Rapid Reading
Dressmaking	Russian
Economic History	Salesmanship
Economic and Social Geography	Soft Furnishing
	Spanish
Economics	Statistics
Electricity	Transport and Distribution
Electronic Computers	Typing

PHYSICAL GEOGRAPHY Made Simple

Richard H. Bryant, B.A., Ph.D.

Made Simple Books
W. H. ALLEN London
A Howard & Wyndham Company

Printed and bound in Great Britain
by Richard Clay (The Chaucer Press), Ltd.,
Bungay, Suffolk
for the publishers W. H. Allen & Company Ltd,
44 Hill Street, London W1X 8LB

ISBN 0 491 01527 5 casebound
ISBN 0 491 01537 2 paperbound

Foreword

Many new and stimulating developments have been apparent in recent years in the study of physical geography. These are not just a question of new information, but reflect the introduction of fresh concepts and frameworks in the subject. Almost inevitably, this has produced an increasing diversity of specialist textbooks that make it difficult for the generalist to keep abreast of all aspects of the discipline. There has also been a tendency for the main parts of physical geography—namely the study of landforms, weather and climate, water, and soils, plants and animals—to evolve their own approaches and objectives.

This book makes an attempt to offer some redress to these trends. It tries to bring together within one volume much of the modern thinking in the subject, and to express these ideas in a simple and concise manner. It adopts the view that physical geography is concerned with the natural environment as a whole, in which its physical and biological components are linked within one vast system. The book also gives more prominence to some of the rather neglected parts of the subject, particularly biogeography, thus modifying the traditional imbalance within physical geography towards the study of landforms.

Students of geography in schools or in colleges should find this book a valuable introduction to modern physical geography. However, no specialist knowledge is assumed on the part of the reader, and the book should therefore be of particular interest to the layman. It is hoped that this book, together with its companion volume, *Economic and Social Geography Made Simple*, will encourage readers to dig deeper into the subject of geography, and some guidance has been given in this respect in the lists of suggested further reading at the end of each chapter.

Many friends and colleagues have given freely of their help and advice during the writing of this book, and it is a pleasure to take this opportunity of thanking them, particularly George Booth, Hazel Faulkner, Kevin O'Reilly and Peter White for their comments on the text; Jean Emberlin for her suggestions concerning the biogeography chapters; Eileen Booth, Alison Pennack and Elizabeth Dawlings for secretarial and technical assistance; and my wife for forbearance throughout.

RICHARD H. BRYANT

Table of Contents

PART FOUR: PHYSICAL GEOGRAPHY AND MAN

PHYSICAL GEOGRAPHY MADE SIMPLE

CHAPTER ONE

INTRODUCTION TO PHYSICAL GEOGRAPHY

Physical geography may be defined as the integrated study of the natural environment on or close to the Earth's surface. **Human geography**, on the other hand, is concerned with man's activities over the surface of the Earth. The nature of the relationship between man and the natural environment is inevitably a complicated one; it varies from place to place, and it has changed through time. In the context of our present awareness of the need to conserve the environment for the best use of all life, including man, the traditional division between the two parts of geography has frequently become one of emphasis rather than of substance.

If physical geography deals with the **natural environment**, what is meant by this expression? First, we should note that, strictly speaking, environment means 'that which surrounds': in its broadest sense this includes all energy and matter capable of influencing man, from the astronomic to the subatomic level. But in practical terms, such subjects as astronomy and nuclear physics are beyond the immediate concern of the physical geographer. He is primarily interested in the visible natural environment, although the basic principles of physics and chemistry are fundamental in explaining how the environment operates. Second, we should be aware that large parts of the Earth's surface cannot now be described as truly **natural**, because of widespread interference by man. In many cases the apparently wild parts of the countryside of Britain and other heavily populated parts of the globe are only semi-natural, and in others, they are highly artificial. An evaluation of man's impact on the natural environment is a theme we shall return to at the end of the book. Nevertheless, in whatever setting, it is important that we make the attempt to understand how natural processes operate in order to appreciate our environment more completely.

Physical geography has been described not so much as a basic science, but as an integration or overview of a number of earth and life sciences which give insight into the nature of man's environment. The question then is, what sciences should be selected to achieve this objective?

First, we need to consider the form of the Earth's relief features. The scientific study of landforms is known as **geomorphology**: this concerns itself not only with the analysis of the shape of landforms, but also with the erosional and depositional processes at work on them and their evolution through time. These aspects of the environment are examined in Part One of the book. Some of the major physiographic features of the Earth, such as mountain chains, continental plains and ocean basins, are a result of internal Earth forces. Hence, certain aspects of **geology**, the study of rocks, are relevant to physical geography. Rock type and structure are also important as variables which influence the effectiveness of wind, rain and weathering processes on landforms. Chapter Two outlines the essential geological background to landform study.

A second major concern of the physical geographer is the atmospheric environment. **Meteorology**, the study of weather processes, together with **climatology**, the analysis of climate or average weather, make up Part Two of this book. The distinction between these two atmospheric sciences is largely arbitrary: the climate of any particular place can only be understood through a knowledge of atmospheric processes.

A third component of physical geography is the study of plant and animal distributions, normally called **biogeography**. The physical geographer needs to be conversant with the basic principles of botany and zoology, and particularly of **ecology**, which studies the relationships between plants and animals and their environment. This is dealt with in Part Three.

These three aspects determine the basic framework of the book, but there are also other disciplines which make significant contributions to the subject. The more important of these include **pedology**, the study of soils, which form an important environmental link between landforms, climate, and plants and animals; **hydrology**, the study of water on the Earth's land areas; and **oceanography**, which covers the study of waves, tides and currents, as well as the biological characteristics of oceans.

The subject clearly embraces a wide range of specialisms and the physical geographer cannot hope to be an expert in them all. But it would be very wrong to imagine that physical geography is simply of potpourri of mappable subjects. It is worth recording that many of the above specialisms, now sciences in their own right, grew out of an original physical geography of a century or more ago. The inevitable trend towards specialism has in no way altered the realities of nature. On the Earth's surface, land, air, water, soils, plants and animals all exist together, and the physical reality of any one place is made up of all these elements. Matter and energy pass continually from one to the other. Although the combination of features may vary from one place to the next, everywhere there exists a tendency towards dynamic balance or equilibrium, in which a change in one of the elements leads to adjustment in the others. The value of physical geography is not only that it studies the important components of the natural environment, but that it concentrates on the connections between them. Modern physical geography tries to interpret the natural environment as a dynamic entity. One way of demonstrating this is to use a **systems** approach, which is outlined later in this chapter. There is a strong requirement today for the 'lateral thought' scientist, who studies the interactions between the various components of the environment, rather than concentrating on a single specialism.

Recent Trends in Physical Geography

This book attempts to be as up to date as possible, tempered it is hoped by an appreciation of some of the difficulties of teaching and learning the subject. Although not all new ideas or discoveries at the research level are worthy of instant assimilation into teaching, many important concepts become buried in an increasing complexity of specialist textbooks. As any teacher of first-year undergraduates will testify, there is frequently a lag of ten or fifteen years between the introduction of new concepts at degree level and these becoming incorporated within school syllabuses. A brief outline is given here of recent trends in the three main fields covered in this book, together with some comment on general changes in the subject.

The study of landforms has undergone a significant change of emphasis in the last twenty years. Any understanding of landforms depends on an appreciation of the relative roles of climate, geology, form, process and time as governing factors. In the first part of the twentieth century, much of geomorphological study placed its emphasis on climate, geology and time. In particular, the subject was dominated by W. M. Davis' **cycle of erosion** (see Chapter Nine), which stressed the evolution of landforms through time, and suggested a classification of landforms based on their stage of development in the cycle. The biggest drawback with this approach was its inability to accommodate effectively the dynamics of present-day processes. In the 1950s and 1960s a strong reaction against Davisian ideas led to their replacement by an emphasis on **process/form** studies, which are concerned with an examination of the relationship between landforms and contemporary processes. The process/form approach can be usefully placed in a systems framework, as illustrated in Chapters Four (slopes) and Five (rivers).

In effect, the Davisian cycle no longer provides an adequate framework for modern geomorphology, and it has therefore not been used in this book as a methodological basis for studying the subject. However, this does not mean that all Davisian ideas are unsound, or that time is not an important factor in landform study. The current emphasis on process studies has had the benefit of making geomorphology much more relevant to the rest of geography, not least in dealing with applied problems (Chapter Twenty-Five). It is more advantageous that we know, for instance, something of the discharge and sediment load of streams, than that they are 'young' or 'mature', as Davis described them.

In the case of weather and climate, there has been not so much a methodological shift, as a tremendous increase in knowledge about the upper layers of the atmosphere. Much of this had come about with the use of satellites and remote-sensing techniques. Weather study has been comprehensively transformed from a two-dimensional view to one in three dimensions. This has had a major impact on climatology. Up to ten years ago climate study was primarily descriptive, simply listing climatic facts for particular regions, or preoccupied with climatic classification. Although these remain legitimate concerns for the geographer, modern climatology also lays much more stress on synoptic or dynamic aspects, such as the analysis of general circulation patterns, airstream characteristics and meso-scale weather systems. Detailed investigations into the global energy budget (Chapter Ten) have helped to revolutionise our view of the general circulation (Chapter Thirteen). Mid-latitude depressions and other meso-scale weather systems are now regarded as important mechanisms in maintaining the circulation rather than accidental disturbances within it. By contrast, fronts are regarded as secondary consequences, rather than the causes of circulation patterns. These new developments allow us a much improved insight into spatial and temporal variations of climate (Chapters Fifteen and Seventeen).

The study of biogeography has for long been a neglected part of physical geography, especially at school level, where syllabuses have tended to be dominated by landform and weather studies. Many physical textbooks confine themselves to descriptions of the major vegetation and soil types of the world. This is a reflection of two characteristics that have symptomised much of traditional biogeography in the past: it has been almost solely concerned with plants, and it has been dominated by the zonal approach and the concept of climatic climax (see Chapter Nineteen). However, in the last decade, the subject has been rejuvenated by a reawakening of interest in plant and animal ecology. Biogeography has begun to focus much more on ecological relationships and processes, especially on energy flow and nutrient cycling. This has served to re-emphasise the use of the **ecosystem** as a fundamental conceptual framework. Hence ecological principles form the basis for the consideration of plants and animals in Part Three of this book. This allows for a much firmer explanation of distributional irregularities, the traditional concern of the geographer rather than the ecologist.

General Trends

As far as physical geography as a whole is concerned it will be apparent from what has already been said that the subject in recent years has become far more **process-orientated**—that is, concerned with explaining the spatial and temporal variations in the environment in terms of the processes operating—rather than simply describing distributions. Far less emphasis is now laid on global classifications of phenomena, particularly those based on climatic indices. This movement towards more rigorous analysis and explanation is part of a trend evident in the whole of geography. Some have called this change a **quantitative revolution** since statistical procedures are now as important a tool as maps in advanced geography, but it might be more accurately described as a theoretical revolution. In brief, it is characterised by a more systematic application of scientific method to the subject. The careful distinction between inductive and deductive reasoning, precise measurement and observation, model building and systems analysis (both further explained below) are some of the manifestations of this new approach.

Another significant general trend is that the subject is becoming increasingly **applied**. This is partly because it is better able to do so as a result of the methodological changes already mentioned, and partly because there is more demand for it to be so in the context of the current desire for better environmental management. The last chapter in this book summarises some of the possible applications of physical geography in this respect. The net result of these trends is that the subject is much more integrated than it has been for some time. In particular the systems approach provides a viable common framework for several of its discrete parts.

Models and Systems

Models

Since the natural environment is so complex, we have to simplify it in some way in order to portray or understand it. Representations of reality are called **models**. We are all familiar with scaled-down models of ships or aircraft which we can physically construct: these are examples of **hardware** models. The term 'model' can also be used to describe **conceptual** idealisations of reality, such as a hypothesis, a law or a theory. The definition of a model proposed by R. J. Chorley and P. Haggett is that it is 'a simplified structuring of reality which represents supposedly significant features or relationships in a generalised form'. In recent years geographers have been making considerable use of models in the application and development of theory: this is a trend which was first apparent in economic geography in the 1950s, but has now spread to physical geography. However, models are by no means new to the subject: a map, a classroom globe and a diagram of a frontal depression, are all examples of models.

Models used in physical geography vary widely in the amount of abstraction of reality in them. Some hardware models, such as large tank models of rivers, estuaries, and coasts, are also **iconic** models, closely imitating the real world in all respects but that of scale. Considerable technical problems are involved in scaling down natural objects in this way. Much more widely used are **analogue** models, which represent the real world by other properties. A kaolin model of a glacier, and a map, are both analogue models. Diagrammatic or mathematical models can also be regarded as analogue models, but involve even more abstraction, replacing objects or forces by symbols or equations. Mathematical models in particular have become very important in geographical research since they can be used to predict changes.

Since they help us to organise and explain data, models are obviously very useful teaching and learning aids, and models of various types are widely used in this book. Models should also help to identify gaps in our knowledge and point the way to further work. However, there are problems in their use which arise from the complexity and diversity of reality. Models are simplifications of reality, and they are often very attractive; there is always the danger that they might be substituted for and accepted as reality. Also, since reality can be simplified in many ways, it is important that any model is considered as but one possible way of viewing the real world. Good models should be capable of being tested against the real world and modified if necessary. Many of the long-standing arguments in the study of physical geography have been about generalised models or laws which are not capable of being proved or disproved.

Systems

With these general points in mind, we can now consider one of the more significant recent developments in physical geography, namely the widespread adoption of models in which the real world is viewed as a vast system or set of interlocking systems. A **system** can be defined as a set of objects or attributes (that is, characteristics of an object, such as size or shape) linked in some relationship. We have already stressed that the natural environment appears to operate as an entity, in which each component has connections with all the other components. It is impossible to build a model which takes in the whole world or even a substantial part of it, so we identify various environmental subsystems within which the connections are fairly strong. Thus weather systems, drainage systems, ecosystems and many others can be described. In analysing these, the systems approach focuses attention on the whole system and the interrelationships within it, rather than on the individual parts.

Examples of systems familiar to us in everyday life include transport networks, the electricity grid system, or the domestic hot-water system of a house. These systems can be modelled symbolically by means of **flow diagrams**, consisting of the objects in the system conventionally represented by symbols, usually box-shaped, and the mass or energy flows in the system represented by lines. Electric circuit diagrams, or the map of the London Underground, are examples of flow diagrams. Figs. 1.1 and 1.2 are very simple flow diagrams; Figs. 5.1 and 10.4 are more complicated examples.

Systems are normally regarded as being of two types: **closed**, in which no energy or matter crosses the external boundaries of the system, as in Fig. 1.1a, and **open**, in which external factors can affect the variables within the system (Fig. 1.1b). Apart from the universe, no natural system is truly closed unless we artificially make it so for the purposes of study. However, the degree of 'openness' of systems varies considerably. The earth and its atmosphere represent a partially open system, exchanging energy with outer space, but to all intent closed to material exchange. Other open systems exchange both mass and energy. For instance, a drainage basin receives energy and mass from precipitation, sunlight and the elevation of the land. These **inputs** pass through the system, doing work on the way to emerge as **outputs** of heat, water and sediment in the sea and atmosphere. A drainage system is typical of many

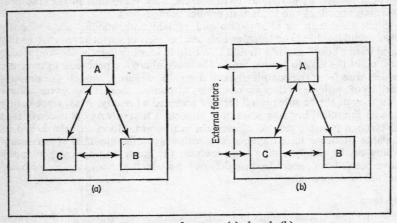

Fig. 1.1. Two types of system: (a) closed; (b) open.

other open systems in physical geography which are called **process-response** systems, because the flow of mass or energy (the process) causes changes of responses in form (shape or arrangement) in the system. However, where plants or animals are involved, as in a forest or a pond (Chapter Nineteen), it is called an ecosystem.

A very significant property of open systems is that the elements within them attempt to adjust themselves to the flow of energy and matter through the system towards a condition of equilibrium or **steady state**. Thus, if we regard a hillslope as an open system (Fig. 4.3), a harmonious relationship will develop over time between the gradient, the infiltration capacity and the size of the sediment particles on the slope. Similarly, in ecosystems, animal populations will adjust closely to plant productivity. The effect of this adjustment is to balance the input of energy and material to the output. However, equilibrium does not mean the system is static, it is performing work all the time, but the opposing forces are balanced or fluctuating about a mean: the state can be alternatively referred to as **dynamic equilibrium**.

A fundamental mechanism in maintaining this state of self-regulation is that of **feedback**. This means that when one of the components in the system changes, perhaps because of some external factor, this leads to a sequence of changes in the other components which eventually affects the first component again. The most common type of relationship is called **negative feedback**, whereby the circuit of changes has the result of damping down the first change. For example, in Fig. 1.2, which is a simplified version of a stream channel system, the increase in stream velocity (A) causes greater erosion (B), thereby increasing channel width (C); but this in turn alters the hydraulic radius of the stream (D) thereby slowing down stream velocity. This kind of negative feedback loop is very common in physical geography, and is the main factor in promoting self-equilibrium and conservatism in natural systems. **Positive feedback**, which is much rarer, occurs when the feedback loop aggravates the original change. For instance, in a glacier system, an increase in velocity leads

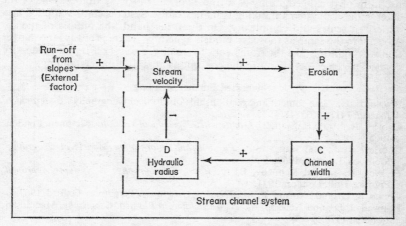

Fig. 1.2. An example of negative feedback.

to an increase in erosion and overdeepening. The overdeepening only serves to accelerate the velocity further. But even here, checks will eventually operate: if the bedrock gradient becomes too steep, this will fundamentally alter the slip-plane field in the ice, such that erosion ceases (Chapter Six). Positive feedback loops in nature usually operate in short bursts of destructive activity, but in the longer term, negative feedback and self-regulation tend to prevail.

Not all systems are in equilibrium, since the speed of response in natural systems to external change varies a great deal. Local weather systems, such as a land-and-sea breeze (Fig. 16.2), exhibit a rapid response to changes in solar radiation through the day. The profile of a beach (Chapter Eight) also responds within a few hours to changes in wave force and direction. On the other hand, bedrock slopes change character very slowly: in Britain and elsewhere, we have glacial features which have been modified little in the 10,000 years of postglacial time and show little equilibrium with present-day processes. Similarly, plant and animal ecosystems often contain features that can only be explained with reference to some past environmental conditions (Chapter Twenty-Two). The timelag between external change and internal adjustment is known as the **relaxation time** of the system. The relaxation time is important in determining the amount of attention we need to pay to the **historical** (time) factor in physical geography. Thus in geomorphology, the landscape adjusts slowly enough for strong historical legacies to remain, and time is an important element to be considered in landforms (Chapter Nine). There are therefore important links between geomorphology and geology. At the other extreme, relaxation times in weather and climate systems are very short, and the approach here is almost entirely in terms of contemporary process. We should note, however, that climatic change itself (Chapter Seventeen) has had an important historical effect on soils, plants, animals and landforms. The value of historical studies is that they add the qualifications we cannot detect in our modern observations.

In summary, we can say that the systems approach is currently proving useful in physical geography as a framework for process studies. One of the chief values of systems thinking is its flexibility. Systems can be applied at a variety of scales and complexity. On the other hand, the pitfalls of the approach are the same as those for models generally. There is always the danger that we might mistake the framework for reality, and set off trying to identify systems *per se* rather than use the concept as an aid to understanding.

Suggested Further Reading

Brown, E. H., 'The content and relationships of Physical Geography', *Geographical Journal*, **141**, pp. 35–48, 1975.

Chorley, R. J., and Haggett, P., *Frontiers in Geographical Teaching*, Methuen, London, 1965.

Chorley, R. J., and Haggett, P., *Physical and Information Models in Geography*, Methuen, London, 1969.

Chorley, R. J., and Kennedy, B. A., *Physical Geography: A Systems Approach*, Prentice-Hall, London, 1971.

Cooke, R. U., and Johnson, R. M., *Trends in Geography*, Pergamon, Oxford, 1969.

Hanwell, J. D., and Newson, M. D., *Techniques in Physical Geography*, Macmillan, London, 1973.

CHAPTER TWO

ROCKS AND RELIEF

Geological considerations play an important role in the development of landforms in two respects. First, landforms are a result not only of erosional and depositional agents acting on the Earth's surface, but also of **tectonic** forces, originating within the Earth. The broad outlines of nearly all the Earth's major relief features, such as mountain ranges, ocean trenches, basins and plateaux, are tectonically formed, although detailed features may be the result of later erosion. Moreover, tectonic forces, in uplifting land, are a major provider of energy into landform systems. Thus it is necessary that we know something of the tectonic nature of the crust of the Earth and of the processes at work in it, particularly in relation to plate tectonics. Second, rocks vary considerably in their resistance to erosion, and it is relevant to consider some of the structural and lithological properties of rocks where they affect the course of landscape development.

Crustal Structure and Movement

The Earth is made up of a series of concentric rock zones; namely, the crust, mantle, outer core and inner core. The **crust** is the main concern here. It varies greatly in thickness and composition; beneath the oceans it is as little as 5 km thick in places, but under some mountain ranges it extends to depths of 70 km. Rocks in the crust fall into two main groups. The ocean basins are predominantly underlain by basaltic rocks, containing much iron and magnesium, and having densities of between 2·8 and 3·3 (water = 1·0). In contrast, the rocks that make up the continents are lighter in colour and weight (densities of *c*. 2·7) and are rich in silicon and aluminium. These rocks are of many kinds, but granitic rocks predominate. The oceanic type of crust appears to extend right round the Earth, passing underneath continental crust, which is confined only to the continents and their shelves (Fig. 2.1).

The difference in density between the two types of rock is thought to explain why continents stand high above ocean basins. The continental areas are not loads supported on a rigid substratum, but instead 'float' on the denser basaltic crust. Each continent is underlain by a root zone of similar light material projecting down into the basaltic zone by an amount proportional to the height of the continental area. The term **isostasy** describes this state of balance. There are important implications to landforms of any disturbance to this structural equilibrium. If material is moved by erosion from an area and deposited on the sea floor, it will involve isostatic adjustment to both areas: a rise in the level of the area subject to erosion and subsidence of the sea floor. Similarly, the addition of weight to a continental area, in the form of ice or a large body of water, for example, will cause the crust to sink slightly. There may be a considerable timelag, perhaps thousands of years, between the

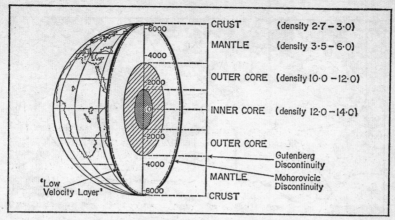

Fig. 2.1. The relative thickness and layering of the crust beneath continents and oceans.

external change and the isostatic response of the crust. Much broad crustal warping today is related to isostasy. Measurable amounts of isostatic uplift are taking place in regions such as Scandinavia and Arctic Canada, which were depressed beneath ice caps during Pleistocene glaciations. However, there are other mechanisms of crustal warping, which will be reviewed below.

Crustal Plates

It is now almost universally accepted by earth scientists that parts of the crust are capable of moving horizontally round the globe, causing the continents slowly to change position in relation to each other. This idea was originally synthesised under the term **continental drift** by A. Wegener in 1915. Considerable opposition to this thesis persisted for many years. In the last fifteen years, however, new discoveries have confirmed his general ideas and have led ultimately to a body of theory which we now call **plate tectonics**. The concepts embodied in this go a long way towards explaining the distribution and origin of many major relief features.

The basic principle of plate tectonics is very simple: the crust is broken into several units or plates (Fig. 2.2). Each plate is made up of the whole thickness of the crust above the mantle, carrying the continental areas embedded in it. Each plate moves as a single independent body. At the mid-oceanic ridges in the Pacific, Atlantic and Indian oceans, plate boundaries are characterised by the creation of new crust from below, and in the process called **sea-floor spreading**, the plates migrate slowly away from these central ridges. Elsewhere, as around the periphery of the Pacific Ocean, plates move past each other or collide. At many zones of collision, one plate overrides the other, the lower plate being absorbed into the mantle, thus maintaining total material balance over the globe, making up for the new crust coming out of the ocean ridges.

At the plate boundaries, major tectonic landforms are created. The areas of collision may be made up of three elements: deep trenches on the ocean floor, some up to 11 km deep, marking areas of plate subduction into the

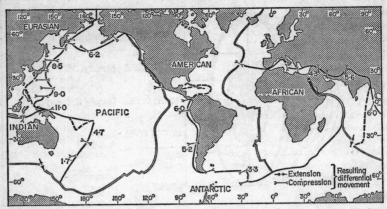

Fig. 2.2. The major plates of the Earth's surface: *double lines*, plates moving apart; *single lines*, plates closing; *dashed lines*, possible minor plates (figures in cm/yr).

mantle; arc-like rows of volcanic islands; and mountain ranges, where the plate appears to be crumpling and thickening. Both mid-ocean ridge systems and plate margins are frequently areas of considerable earthquake activity and volcanism.

Earth Movements

Movements of the crustal plates cause pressure and tensions to build up on rocks, in many cases leading to deformation of the land surface. The general term **diastrophism** is sometimes applied to the bending, folding, warping and fracturing of the crust. It is important to distinguish between several types of movement and their results.

On a broad scale, earth movements may be divided into two types. **Epeirogenic** movements are those involving forces acting along a radius from the Earth's centre to the surface, and are characterised by large-scale uplift or submergence of land areas. The rocks involved in epeirogenic movements are typically warped and tilted rather than intensively folded or fractured. The second type of earth movements are those generated by forces acting at a tangent to the surface of the Earth, as primarily involved in plate tectonics. Where such disturbances have been responsible for the formation of the great fold mountain ranges of the world, they are referred to as **orogenic** movements. The creation of very complex fold structures, as sometimes involved in orogenesis, is called **tectogenesis** by some authors.

From the point of view of landform development, the difference between epeirogenic and orogenic crustal movements can be quite striking, and this is illustrated with reference to fold mountains and block mountains later in the chapter. In orogenic movements, structurally identifiable units are usually difficult to recognise, but the results of epeirogenic movements may be clearly defined in the relief.

Rocks vary considerably in their behaviour to earth movements. Under surface conditions, rocks are brittle and fracture when subjected to stress and pressure, causing **faulting**. More deeply buried rocks, subject to higher temperatures and pressure, are relatively plastic and may respond to stress by **folding** rather than fracturing. The effects of folding and faulting on the disposition of strata are considered later in the chapter.

Earthquakes are the most prominent evidence of present-day earth movements, and are the result of deformation in the crust, which finally ruptures abruptly. Earthquakes are important in landform studies because they can trigger off catastrophic events in erosion processes. These include large-scale landslides and mudflows, as in Peru in 1970. A number of spectacular glacier surges have been observed in Alaska, and some are thought to be associated with earthquakes. **Tsunamis** are seismic sea waves which can arrive at coasts with great force, causing considerable damage and shoreline changes, as in recent decades in the Hawaiian islands.

Major Tectonic Landforms

Mountain Chains

Examination of a relief map of the world will reveal a number of major mountain chains arranged in long linear arcs. Notable among these are the Alpine–Himalayan system and the Circum–Pacific system comprising the Andes, the Rockies and island chains of east Asia. These systems each consist of series of fold structures aligned in a roughly parallel manner. They appear to have formed as a result of the movement of plates (Fig. 2.3a) at different times in the earth's geological history.

These episodes may mark the closure of former oceans. Nearly all fold mountain systems involve considerable thicknesses of sedimentary rocks that have been crumpled and folded. Accumulations of over 8 km of sediment are involved in the Appalachian mountain system, for instance. The sedimentary rocks here and elsewhere characteristically exhibit signs of being

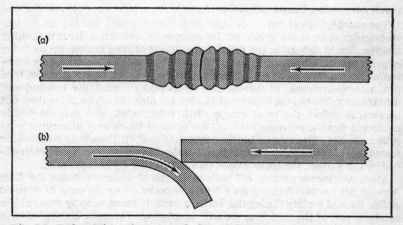

Fig. 2.3. Deformation where crustal plates collide: (a) thickening of plates in collision, forming a mountain range; (b) oceanic plate bending under a thicker continental plate, creating an oceanic trench.

deposited near continental edges and consist of sandstone-shale alternations sometimes known as **flysch**. The thicknesses of material involved were once thought to indicate that the sediments accumulated in slowly subsiding areas of shallow water termed **geosynclines**. However, now that it is realised that turbidity currents can deposit sand in deep water, the thick sedimentary sequences involved in mountains are regarded as large wedges of material which accumulated on continental slopes at the outer edge of continental shelves. The alternations in these sediments are important to landforms in the features of differential erosion that they may produce.

The transformation of these sediments into mountains may have been the result of both compressional forces and isostatic uplift. In some mountain systems, the uplift also appears to be associated with the intrusion of large bodies of igneous rock. The granitisation of the root zone may lead to a reduced rock density and increased volume, which causes the whole orogenic belt to rise. However, by whatever mechanism the chain is formed, gravitational sliding of the upper central areas often seems to have accompanied uplift, creating nappe structures or complex folds.

Many of the world's largest mountain chains exist beneath the sea. Some of these are revealed as **island arcs**, as in the West Indies, and in the west and south-west Pacific ocean. Associated with them are deep oceanic **trenches** on the convex sides of the arc. These features are a direct result of the movement of crustal plates. The island arcs represent the upper parts of anticlines of compressional origin formed by the buckling of a plate as it overrides another one; the ocean deep marks the site of the downward plunging of the lower plate into the mantle (Fig. 2.3b). The mid-oceanic ridges form the longest mountain chains. The mid-Atlantic ridge rises 3 km above the floor of the Atlantic; it is connected with the Indian Ocean ridge, and thence with the Pacific–Antarctic ridge, resulting in a continuous system some 40,000 km in length.

Block Mountains, Basins and Rifts

The morphological results of large-scale warping and faulting are no less spectacular than those produced by orogenic movements. Recent faulting is often clearly defined in the relief. Large tracts of land broken up by faults of great vertical displacement may form **block mountains**, separated by intervening basins. The Basin and Range country of western North America is of this nature, a mosaic of sub-parallel faults and differentially uplifted and tilted blocks. Within the individual blocks, the attitude of the strata does not necessarily reflect the most recent earth movements, and may be highly contorted from a previous episode. An uplifted block may alternatively be called a **horst**; a downthrust block, a **graben** (Fig. 2.4). These may be small, or form large elongated rift valleys. Horst-and-graben structure is well exemplified by the Vosges, the Black Forest (horsts) and the Rhine rift valley.

Most continents possess **rift valleys**, the most extensive being the East African rift system. Present-day rift valleys occur along the crest of tectonic arches formed mainly during the Tertiary period; some, such as those at the southern end of the Red Sea, are still developing. There are several ways in which a rift valley may form. The usual explanation is that the rock floor sinks between two inward-facing fault scarps, but this is rare. More commonly, there exist a number of faults on each side of the valley, sometimes arranged

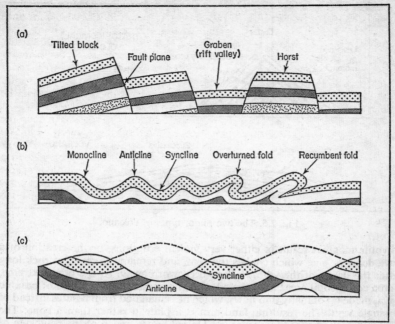

Fig. 2.4. Diastrophism at the Earth's surface: (a) block faulting (uneroded); (b) simple folding (uneroded); (c) folding showing inversion of relief.

en echelon. In other cases, no fault might be apparent at all at the surface, and instead, strata on either side are steeply downwarped towards the floor of the valley. Volcanoes are frequently associated with rift valleys, taking advantage of the crustal weaknesses set up by faulting.

Volcanoes

Volcanic landforms are very diverse, mainly as a result of chemical differences in the magma which feeds different sites. This not only creates different structural types of volcano, but also results in landforms that vary greatly in their resistance to erosion.

Two major types of volcano are generally recognised. Outpourings of very fluid basaltic lava are usually accompanied by little violent eruptive activity. Individual lava flows are normally only a few feet thick, but over a long period of time, repeated flows build up **shield volcanoes** (Fig. 2.5), as exemplified by those of the island of Hawaii. Here, Mauna Loa and Mauna Kea are made up of thousands of flows to rise 9 km above the ocean floor. Cooler and more viscous andesitic lavas build volcanoes that are characterised by explosive eruptions and the ejection of a wide range of pyroclastic material (ash, cinders and rock fragments). Cone-shaped volcanoes result, otherwise known as **strato-volcanoes**.

Other types of volcano sometimes recognised include the **composite type**, consisting of several vents and parasitic cones. Extremely viscous acid or

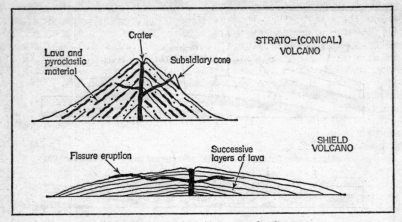

Fig. 2.5. The two major types of volcano.

rhyolite magmas result in either very violent eruptions, or the creation of a **lava dome** or **plug** which chokes the vent, and remains as resistant rock long after the flanks of the volcano have been worn away. At the other extreme, some continental areas are covered with enormous accumulations of basaltic lava, made up of many thin flows, which has emanated from **fissures** instead of a single vent. The resulting landform is a plateau rather than a cone. The Columbia plateau of the north-west United States, much of the peninsula of India and portions of South Africa are of this nature.

Most active volcanoes today are concentrated in several well-defined zones, mainly along plate margins. The best known is the Pacific 'Ring of Fire', largely made up of explosive andesitic volcanoes. The mid-oceanic ridge systems are entirely volcanic, as are many individual oceanic islands. Some chains of volcanic islands—for example, the Hawaiian group—show a progression of increasing age away from the most recently active vent. This sequence is attributed to the gradual passage of the oceanic crust over a so-called **hot-spot** in the mantle underneath.

Rock Structures and Landforms

When tectonic activity ceases or becomes very slow, external weathering and erosion forces gradually become the dominant factors in the sculpturing of the landscape. The influence of geology on the landscape passes to a more detailed level where even minor variations in the lithology and structure of rocks may have an influence on the landforms.

Joint Structures

All rocks develop joints as they consolidate and crack under the stresses set up by cooling or pressure changes. Both intrusive and extrusive igneous rocks develop columnar joints in response to contraction while cooling, well exemplified by basalt. Sedimentary rocks usually develop joints at right angles to their bedding planes. Some jointing patterns are systematic (regular), while non-systematic joints are generally curved fractures which do not cross other joints. Joint directions have a profound influence in guiding the course of erosional processes, which becomes reflected in the alignments in landforms. For instance, in many valley floors, streams follow joint directions, particularly in jointed igneous rocks and on flat-lying sedimentary rocks. A striking influence on drainage lines in the upper Mississippi has been noted. Studies on East Yorkshire rivers in England also have shown that joint directions in Millstone Grit and other rocks strongly influence valley alignment. At a more detailed level, cross jointing in granite is fundamental to the shape of tors (see Fig. 3.1).

It is of considerable significance to landform studies that joint frequency in most rocks apparently decreases with depth, and that many joints are formed near the surface. It seems that many joints only open up when erosion occurs. This process is known as **unloading** or the **pressure-release mechanism**. Some types of unloading result in the formation of joints in accordance with original lines of weakness in the rock, whereas in other cases, **sheeting** may occur, creating joints parallel to the existing ground surface, independent of all primary structures. Sheeting particularly affects granites and quartzites.

The significance of joints in facilitating erosion is therefore complicated by the fact that the joints themselves seem to depend on erosion. This relationship may be self-perpetuating, since there is probably a tendency for distinct joint-orientated landforms, such as domes, canyons and cliffs, to persist. This is an example of feedback in geomorphology, whereby the form (the cliff) controls the process of unloading, which in turn determines the form.

Faulting

As used earlier in this chapter, a fault is a fracture in rock along which rocks have been relatively displaced. The amount of displacement, which can be horizontal or vertical, may be very small and hardly distinguishable from a joint, or it may be many tens of kilometres. It is possible to describe faults by several geometric or genetic criteria. For our purposes, three types may be distinguished. In a **normal fault** (Fig. 2.6), usually the result of tensional forces,

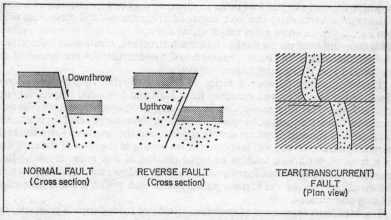

Fig. 2.6. Types of fault.

the inclination of the fault plane and the direction of downthrow are either both to the left or both to the right. In a **reverse fault**, resulting from compressional forces, the beds on one side of the fault plane are thrust over the other. A **tear** or **lateral fault** is one in which rocks are displaced horizontally along a line of fracture, with little or no vertical movement. The amount of vertical displacement, if any, on a fault is referred to as the **throw** of the fault; the amount of lateral displacement is known as the **heave**.

The surface expression of a fault depends on the hardness of the rocks involved and the size of the displacement. Normal faulting in competent rocks produces a **fault scarp**, a sharply defined cliff-like feature. With the passage of time, the original fault scarp may become subdued, but the fault plane may have a control on the landscape for long afterwards, especially if rocks of different hardness have been brought together on either side of the fault. Such features are called **fault-line scarps**. Elevations on either side of a fault-line scarp may not correspond to the direction of original vertical movement if less resistant material is on the side of the original relative uplift.

Faults frequently tend to occur in zones, where the affected strata may form a dislocated belt perhaps several hundred metres wide, as on parts of the San Andreas fault. These zones may be characterised by crushed or shattered rock known as **fault breccia**, and present situations which favour linear erosion. The fjords of western Norway largely coincide with faulted and crushed zones, indicating that fluvial and subsequent glacial erosion were largely structurally controlled. In the block-faulted country described earlier, faults may stand out boldly, but equally, a large number of faults exist without any trace in the landform.

Folding

The nature of folded structures becomes particularly significant in landform studies where several contrasting lithologies are involved. Much depends on the degree of dip. In a simple series of folds, **anticlines** (upfolds) may be distinguished from **synclines** (downfolds). The folding may be symmetrical, or where the limbs of the fold differ in dip, they may be asymmetrical. Degrees of increasing asymmetry can be recognised in **overturned** and **recumbent folds** (see Fig. 2.4b). In some cases major anticlines and synclines may have minor folds superimposed on the flanks of the main structure, creating anticlinoriums and synclinoriums. In Britain, examples of these structures are provided by the Weald and Hampshire basins respectively.

The relationship between a series of folds and relief may be considered **direct** where anticlines and synclines form high and low relief respectively (as in Fig. 2.4b). The Jura mountains in northern Switzerland are a well-known example. Quite frequently, however, there is an **inversion of relief**, in which the axial lines of the anticlines are deeply eroded, and the synclines form the high ground. One explanation of this cites the fracturing of the apex of the anticline as a zone of weakness, leading to rapid erosion of this zone. It seems that erosional rates may be sufficiently effective during slow rates of folding so that the synclines become the highest points in the area within a short time after folding commences.

The landform expression of individual limbs of folds depends primarily on two factors: the inclination of the strata, and the arrangement of varying lithologies. One significant effect of inclination is that the relative erodibility of a rock is expressed most accurately in the landscape when dips are vertical, and least accurately when dips are low. In the latter case, a resistant **cap-rock** may protect weaker underlying strata. Some examples of this in relation to scarplands are illustrated in Chapter Four (Fig. 4.7). Moderate dips accentuate the importance of gravity in erosional processes, particularly on sea cliffs and other similar situations where the lower end of a tilted block of strata is being eroded away, and the dips are seaward. The effect of varying lithologies in tilted strata is to produce examples of **differential erosion**, of which the alternations of scarp and vale in the scarplands of south-east England are a well-known example.

Rock Type and Landforms

Resistance to Erosion

The resistance of an individual rock type depends on a large number of variables, including mineralogical composition, grain size, the nature of the cementing agent, individual grains and permeability. These factors are particularly relevant in relation to the effectiveness of weathering, and the response of a number of rock types to weathering is considered in the next chapter. But in all landscapes, the most important factor is not so much the absolute resistance of a rock, but its relative resistance in relation to the strata around it. It is this which creates the pattern of high and low relief. For instance, chalk is a relatively hard rock in relation to adjacent strata in south-east England, but in Northern Ireland, it is soft compared to surrounding basalts.

The detailed effects of variations in relative resistance of a rock to erosion may be illustrated with reference to igneous intrusions. **Dykes** are a common form of intrusion which are usually **discordant** to the country rock. They are vertical or near-vertical structures, varying in thickness from a few centimetres up to about fifty metres in thickness. The relief effect of dykes varies greatly. Where they are harder than the surrounding country rock, they will tend to form a wall-like ridge (Fig. 2.7a), sometimes traceable for several kilometres across country. Where dykes are weaker than surrounding rocks, they will form a trough (Fig. 2.7b). Sometimes the rock in immediate contact with the dyke will have become metamorphosed so that it forms either raised (Fig. 2.7c) or channelled (Fig. 2.7d) margins to the dyke. Contrary to what is generally assumed, dykes in Britain rarely form continuous wall-like features, since they are often intruded into other igneous, or metamorphic rocks, of greater resistance than themselves.

Sills, on the other hand, generally act as harder members of the rock series into which they have been intruded. Unlike dykes they are intrusions which are **concordant** with the bedding planes of the country rock, and they are usually intruded into weaker sedimentary strata. In addition, they are nearly all horizontal in attitude. Thus, although they are composed of the same range of materials as dykes, they are often bold relief formers, creating ledges or scarps in many places. Whin Sill in northern England is a well-known example. It is intruded into Carboniferous sediments and for most of its length forms a bold escarpment or small craggy hills. Where crossed by streams its presence is often indicated by waterfalls.

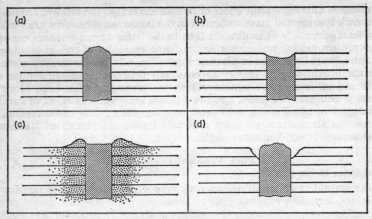

Fig. 2.7. Example of the varying relative resistance of rocks: (a) dyke more resistant than country rock; (b) dyke less resistant; (c) rocks metamorphosed by contact with the dyke stand higher than the dyke itself; (d) dyke with channelled margins.

Distinctive Lithologies

It is sometimes asserted in landform studies that distinctive lithologies create distinctive landscapes. However, it is hoped that this chapter has shown that there are certain qualifications to this idea. Rock types form different kinds of relief in different structural situations. Moreover, a single rock type may not always behave the same way to weathering in a variety of climatic situations (Chapter Three). Thus it is not possible to make broad generalisations about the effects of a large number of specific rock types on relief, and distinctive rock-dominated landscapes are relatively few. Two examples are worthy of attention, however, those of limestone and granite. Each illustrates some of the variations in landforms caused by local factors.

Limestone Relief

Two related facts account for many of the noteworthy features of limestone relief. First, limestone rocks are predominantly composed of varying proportions of calcium and magnesium carbonate, which in the process of carbonation (see page 25), are soluble in rainwater. Second, limestone strata are **permeable**, that is, they allow the transmission of water. The permeability is of two kinds: water passes through the pore spaces in the rock (the property of **porosity**), and also through the lines of weakness, such as bedding planes, joints and faults, which are widened by concentrated water solution. Nearly all limestone landscapes are thus characterised by a high degree of subsurface drainage. However, relative variations in the two types of permeability, dependent in turn on the chemical composition and structural properties of the rock, give rise to a wide range of relief.

On hard massive limestones, where joints and bedding planes are well developed, **karst** scenery may be found. This is typified by large-scale solution features such as sinks, underground caverns and steep-sided gorges. Horizontally bedded strata create extensive limestone **pavements** with their own distinctive microrelief of widened vertical joint planes. The type area for this relief lies in the Dolomites of Yugoslavia, but good examples exist in Britain on the Carboniferous limestones of the Pennines and the Mendips.

The rounded relief forms developed on **chalk**, a porous and relatively pure carbonate limestone, stand in marked contrast to the angular features of karst. Chalk has a much smaller-scale and more weakly developed jointing pattern than massive limestones, and has greater scarcity of underground solution features. Instead, solution at or near the surface is relatively high.

Although in Britain part of the contrast between the two types of limestone may be attributed to the fact that many massive limestone areas have been glaciated, whereas the chalk of southern England has not, it is clear that the range of relief developed on limestones can be vast. The properties that determine this are largely hydrogeological. In view of this range, the term 'karst' should be used in a restricted sense, and not to cover all types of limestone scenery.

Granite Relief

Granite is popularly considered a 'hard' rock, and therefore a distinctive relief former. It originates in **batholiths**, large igneous intrusions which cool at depth, but which are quite commonly exposed, especially in the core of mountain areas. An ideal batholith might be circular or oval in form, its original domed form imparting some distinctiveness to granite relief. But many batholiths are found in ancient worn-down shields of crystalline rocks, as in the Laurentian and Fennoscandian shields. Here, and in the Highlands of Scotland, numerous granitic areas have little obvious effect on the relief. On the other hand, where granites are intruded into sedimentary rocks, as in the southern uplands of Scotland, the overall effect is much more obvious.

The effects of jointing in granite can be most marked. The joint pattern may strongly control the detailed morphology of a cliff face, as in the cuboidal joint pattern at Land's End. The development of sheet jointing through unloading is common, maintaining domed and rounded forms where bare rock is exposed. Glaciation has a notable effect on revealing joint patterns, as in Yosemite National Park, California. In the humid tropics, joints may facilitate deep weathering and then become masked by its effects.

A special feature of many granite landscapes is the presence of **tors**, small rocky hills. In detail, their form is very much controlled by jointing (Fig. 3.1). Although there is no general agreement as to the precise processes involved, they are probably the product of a period of deep chemical weathering, guided by the joint pattern, followed by the later removal of the weathering products after a change of either base level or climate.

Some authorities regard granite relief as **zonal**—that is, it varies fundamentally with climatic regime. Thus in the humid tropics, weathering produces a deep mantle of fine debris covering the solid rock; in tropical areas with a wet and a dry season, variously named granite domes may form by the massive exfoliation of material; and at the other extreme in cold regions, freeze–thaw action reduces granite to masses of angular debris. However, there are many other considerations to be taken into account which make this relationship difficult to justify, such as the complications of climatic change on landforms, and the effects of local variations in the texture of granite. As with limestone, there is an almost bewildering variety of forms on granite, such that simple generalisations about the effect of rock-type on relief require considerable qualification.

Suggested Further Reading

Doornkamp, J. C., 'Rift valleys and recent tectonics', in *The Unquiet Landscape*, edited by Brunsden, D., and Doornkamp, J. C., IPC Magazines, London, 1974.

Francis, P., 'Volcanoes', in *The Unquiet Landscape*, op. cit.

Le Pichon, X., *et al.*, *Plate Tectonics*, Elsevier, Amsterdam, 1973.

Sparks, B. W., *Rocks and Relief*, Longman, London, 1971.

Tricart, J., *Structural Geomorphology*, Longman, London, 1974.

CHAPTER THREE

WEATHERING

Weathering may be described as the disintegration or decomposition of rocks *in situ* by natural agents at, or near, the surface of the Earth. Weathering changes hard massive rock into more fine material. For this reason, weathering is often described as the first essential phase in the denudation of the landscape, as it prepares rock materials for transportation by the other agents of land erosion, including mass movement of material down slopes. Rock weathering is also an important prerequisite to the formation of soils (Chapter Eighteen). A typical soil profile will reveal a gradation of weathering downwards to parent rock.

Although the importance of weathering in the preparation of land surfaces is widely recognised, many weathering processes are only understood in outline. It is convenient to explain individual weathering processes in isolation, but in reality several processes usually combine to cause rock weathering. Two general types of weathering are normally recognised: **physical** (mechanical) **weathering** involves rock disintegration without any change in the chemical constituents of the rock; in **chemical weathering**, on the other hand, some or all of the minerals in the rocks suffer decay or alteration by such agents as water, oxygen, carbon or various organic acids.

Physical Weathering

The main factors responsible for physical weathering are temperature changes; the crystallisation of water into ice or other crystal growth; the pressure-release mechanism; and the mechanical action of animals and plants.

The **thermal expansion** of rock has long been cited as an important cause of rock cracking and disintegration. Many travellers in the arid tropics, where daytime temperatures are at the most extreme, have reported hearing in the evening or night, sounds like rifle shots which they attribute to the cracking of rocks as they contract. The theory is that rocks are poor conductors of heat; given strong diurnal heating, the outer layers of the rock warm up considerably, but do not transmit heat to the inner layers. This should lead to the setting up of stresses in the rock, causing fracturing parallel to the surface. This process has been termed **exfoliation**, although we must note that this word should be more correctly applied to the pressure-release mechanism described below.

Considerable doubt has been expressed by many geomorphologists as to the reality of this simple thermal expansion process. In the Egyptian desert near Cairo, where granite columns have fallen on their sides, the parts fully exposed to the full blast of the sun are surprisingly the least weathered, whereas the shaded areas, in contact with moisture in the sand, are much more decayed. Experimental work has also shown that heating and cooling do not of themselves produce detectable weathering. Attempts have been made to simulate in the laboratory the equivalent of many years of fluctuating temperatures in deserts. Little disintegration was recorded under dry conditions, but in the presence of moisture decomposition became rapid. One of the important conclusions from these experiments is that chemical weathering, involving water, is probably quite important in deserts.

The explanation of the large-scale mechanical exfoliation of rock may lie in the availability of parallel curved structures created by the pressure-release (unloading) mechanism described in the last chapter. Both mechanical and chemical weathering processes further weaken the joints, the layers thereby peeling off in sheets. Hence it is probably best to conclude that chemical weathering and pressure release ally with temperature changes to produce rock disintegration in arid areas.

Ice crystal growth operates as a physical weathering agency in a rather different manner. When water turns to ice, it undergoes a nine per cent increase in volume. In the subarctic zones of the world, or at high elevations, temperatures frequently oscillate about the freezing point, and the consequent alternations of water between its liquid and solid state cause pressures in rock crevices sufficiently powerful to disrupt many rocks. Rapid disintegration of such rocks is achieved, creating a mass of frost-shattered debris. This process is sometimes referred to as **freeze-thaw disintegration**. The process appears to be most effective on well-bedded or jointed rocks, where a large number of suitable crevices are likely to exist. Some porous rocks, such as chalk, are also liable to be considerably affected, since there are plenty of pore spaces in which water may be retained. The same applies to moisture-holding clays. However, free-draining sandstones are not so susceptible to physical weathering of this sort.

Freeze-thaw rock disintegration is widespread nowadays in mountain ranges such as the Alps and Rockies, as witnessed by the abundance of **screes**. These are accumulations of angular debris at the foot of frost-shattered cliffs. In Britain, many fossil or semi-active screes are found, indicative of the more widespread nature of the process in the last Ice Age.

The crystallisation of salts can cause the disintegration of rocks under certain conditions. A number of salts, such as sodium chloride (common salt), calcium sulphate (gypsum) and sodium carbonate may enter rocks in dissolved form. On drying and crystallisation they expand and set up a disruptive effect. Crystallisation has been observed to occur against pressures as great as 47 bars. The tensile strength of many rocks is as low as 20 bars, thus crystal growth can inevitably cause splitting. It is likely that hydration (see below) may also be involved when crystallisation takes place.

Salt crystallisation produces **cavernous weathering**, of which the small-scale honeycombing of rock surfaces is a well-known example. Older explanations for honeycombing, such as wind abrasion or chemical weathering, are nowadays regarded as of lesser importance.

The role of **plants and animals** as agents of physical weathering is fairly limited. Tree roots can occasionally be shown to have forced apart adjacent blocks of rock. Worms, rabbits and other burrowing creatures may help in the excavation of partially weathered fragments of rock. However, organisms do play an important part in soil formation, continuing the process of weathering a stage further (see Chapter Eighteen).

Chemical Weathering

The chemical weathering of rocks is accomplished in a variety of ways. Although no rock-forming mineral is absolutely chemically inert, some minerals are far more readily altered than others—particularly olivine and augite, which occur widely in basalt. Even quartz, which is usually regarded as one of the most stable of minerals, is slightly soluble in water, and as with many other minerals, it is more soluble in saline waters. Four or five main processes may be recognised.

A few minerals such as rock salt are significantly soluble in water. But these are not common constituents of rocks and simple **solution** is not a common weathering process. What many people mistake for solution is the process of **carbonation**. Many minerals are soluble in rainwater, which contains carbon dioxide and acts as a weak carbonic acid. This is particularly important in the decomposition of limestones; the rainwater converts the calcium carbonate into calcium bicarbonate, which is soluble and can be taken away in the groundwater:

$$CaCO_3 + H_2O + CO_2 \rightarrow Ca(HCO_3)_2$$
(calcium carbonate)   (calcium bicarbonate)

The acid of rainwater is supplemented by other acids, particularly those derived from plants and peat.

Hydration is a process by which certain types of mineral expand as they take up water, causing additional stresses in the rock. For instance, calcium sulphate occurs in both hydrated and unhydrated states and is known as gypsum and anhydrate respectively. Many of the decomposition products of rock-forming minerals are subject to hydration, thereby accelerating the disintegration of the rock. In itself, hydration is a mechanical effect, but it occurs intimately with hydrolosis in such a manner that it is difficult to draw any hard and fast line here between mechanical and chemical weathering.

Hydrolysis is a reaction very important in the decomposition of felspars, which are present in many igneous rocks, leading to their disintegration down to basic clay minerals. The reaction can be expressed in words as: crystalline felspathic rocks plus water → clay, sand and carbonates in solution. The process is accelerated by the presence of carbonic acid in rainwater.

The process of **oxidation** occurs when minerals in freshly exposed rocks take up additional oxygen. Deep-buried clays are often blue or grey in colour as long as air is excluded from them, but on exposure they are oxidised and turn red or brown as ferric compounds are formed. To put it simply, they rust. This phenomenon is well seen in exposures of London Clay in southeast England: road cuttings and pit faces change colour even after a few weeks.

Plants may help to promote chemical weathering by providing humic acids. In addition, the process of carbonation in soil and weathered rock is accelerated by the presence of many small animals, which through respiration can increase concentrations of carbon dioxide in the ground to many times the atmospheric content. An interesting specialised case of weathering is the action of lichens in extracting iron from certain rocks, especially granites, and concentrating it at the surface.

Factors Affecting Weathering

The weathering processes we have been looking at are universal in the sense that they can potentially take place in nearly all environments. However, the degree of weathering and the actual combination of processes can vary considerably from place to place. The two main factors which ultimately determine the type and course of weathering are the rock type itself and the climate. Some effects are direct, others operate through vegetation and soil. Geology and climate combine in controlling the availability of moisture in the weathering zone; water is vital to nearly all types of weathering.

Rock Type

Dealing first with the **igneous group,** on the whole, basic rocks such as basalt, containing much olivine, augite and lime-rich felspar, weather more rapidly than acid rocks such as granite. Another general rule is that dark-coloured minerals are more susceptible to chemical weathering than light-coloured minerals. As far as texture is concerned, various crystals in an igneous rock rarely weather at the same rate, and therefore we can expect a coarse-grained granite to weather more quickly than a fine-grained granite of the same composition. This is because in the coarser rock the weathering of one constituent will have a proportionally larger effect in weathering the whole fabric of the rock; in addition, fine-grained granites often possess a better interlocking structure between their crystals.

Structural weaknesses in igneous rocks are principally joint planes; the number of these determines the number of avenues open to weathering agencies. Intrusive igneous rocks tend to develop regular horizontal and vertical joints, creating angular blocks which become rounded during the course of weathering. This type of change is sometimes known as **spheroidal weathering,** and is well seen on the blocks which make up the granite tors of Dartmoor. Much of this weathering may be accomplished below the surface; Fig. 3.1 illustrates deep spheroidal weathering of differentially jointed rock. The well-defined polygonal columns of basalt also present large surface areas to weathering attack.

Sedimentary rocks vary considerably in their resistance to weathering. Conglomerates and sandstones usually consist of pebbles or smaller particles of quartz, which is not normally affected by chemical weathering. Silica, of which quartz is one form, is extremely durable, as witnessed by the resistance of flint, which survives as beach or river sediment long after the surrounding chalk has been disintegrated. Conglomerates and sandstones are susceptible to weathering only in the material which cements the grains together. The most durable rocks are therefore silica-cemented sandstones called quartzites, which are probably the most resistant rock on the Earth's surface. On the other hand, if the cement is calcium carbonate, it is liable to carbonation, and equally if the cement consists of iron oxides it may be subject to the process of oxidation.

Clays and shales are not generally susceptible to chemical weathering except through the impurities they contain, such as ferrous oxides. However, clays and shales are usually bedded or laminated and contain some water, thus being susceptible to freeze–thaw action and other physical processes, and readily reduced to their original fine-grained texture. Calcareous rocks—

that is, the limestones—are very susceptible to the action of carbon dioxide and humic acids. They are usually very permeable, since the well-pronounced jointing systems present a large area available for weathering.

Of the metamorphic rocks, granulites (coarse quartz-felspar rocks) and gneisses weather slowly in much the same way as granite, as their composition is somewhat similar. Schists, containing layers of mica, are more susceptible to weathering, as the layers allow weathering agents to penetrate more readily. Similarly, in slates the cleavage helps weathering to cause flaking of the rock.

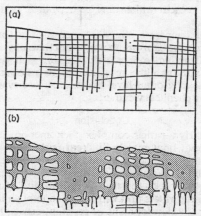

Fig. 3.1. Stages in the deep weathering of differentially jointed rock.

Climate

Climate affects the relative importance of the different weathering processes by controlling rates of operation. Broadly speaking, chemical weathering predominates in tropical or temperate humid climates, and physical weathering in cold or dry regions. But this statement needs some qualification, which may be illustrated in the context of four climatic regions.

In **humid tropical** areas, temperatures are consistently high, and moisture and humic acids derived from decaying vegetation are everywhere abundant. The rate of most chemical reactions increases with rises in temperature and moisture, and undoubtedly chemical weathering is more pronounced here than physical weathering. This seems to be borne out by the great depths of rotted rock found in many places, but no precise data are available on how long it took to create such depths. Whereas outside the tropics typical end-products of weathering are sand and clay, in the humid tropics even these are unstable, and this leads to the development of residual deposits of laterite and bauxite, rich in iron and aluminium respectively.

By contrast, the dominant weathering process in **cold regions** is freeze–thaw action, producing great piles of shattered angular debris. But it has also been suggested that chemical weathering may be important, particularly as carbon dioxide is about twice as soluble at 0°C as at 20°C. Snow-banks have a high concentration of carbon dioxide and this probably has some effect in weathering out the hollow in which the snow-bank lies. Elsewhere, however, since vegetation cover is so sparse, available carbon dioxide in the soil is probably rather low.

We have seen that chemical weathering may be more important in **deserts** than one would perhaps imagine. This not only applies to the so-called exfoliation process, but also in the further reduction of rocks to a sandy debris roughly corresponding to the individual crystals in the rock. However low the relative humidity in deserts, it is never completely dry and there is always some water vapour present which may be deposited as dew at night on the rapidly cooling rocks.

Finally, in **humid temperate** climates, such as that of the British Isles, chemical weathering is probably more important than physical, except possibly at elevations over 600 m, where freeze–thaw action is more effective than elsewhere. In Britain, although the total rainfall is not very great, rocks are almost always moist at ground level because of low evaporation rates, and this helps to promote chemical weathering.

Conclusion

Weathering is clearly a rather complex phenomenon. It is very difficult to derive firm rules about how exactly different rocks will weather in different climatic environments, as much depends on minor differences in lithology and moisture availability. We have noted that the chief significance of weathering is that it prepares rocks for transportation and erosion, and without preliminary weathering, rivers and wind can accomplish little. Weathering is also responsible for the development of resistant crusts, hardpans and horizons in soils. These can render a surface impenetrable, with significant consequences for run-off and erosion. Equally, however, weathering itself is to a large extent dependent on the operation of other processes to transport away the products of weathering and re-expose fresh rock, otherwise weathering will necessarily slow down. Hence we find that on a hillslope the processes of weathering, transport and erosion are very much mutually interdependent and form one unified system.

Suggested Further Reading

Carroll, D., *Rock Weathering*, Plenum Press, New York, 1970.
Ollier, C. D., *Weathering*, Oliver & Boyd, Edinburgh, 1969.
Sparks, B. W., *Rocks and Relief*, Longman, London, 1970.

CHAPTER FOUR

SLOPES

In everyday conversation, most of us use the word 'slope' in a fairly restricted sense to refer to an area of land or a line that makes a definite angle to the horizontal. In geomorphology, a much wider view is sometimes taken, in which a slope includes any facet of the solid land-surface, be it flat or otherwise. The entire landscape is made up of slopes, and on this basis one can say that geomorphology is entirely concerned with slopes created in a wide range of subaerial, submarine and subglacial environments. **Subaerial** slopes include **aggradational** slopes created by river sediments (alluvium), rainwash (colluvium) and rockfall (talus). **Degradational** slopes include cliffs and escarpments, watersheds, and surfaces directly eroded by streams. In practice, however, 'slope studies' in geomorphology are primarily concerned with **hillslopes**—that is, the slopes connecting interfluve crests (hilltops) with river channels in valley bottoms. In this way we can regard slopes as dynamically integrating the whole subaerial landscape, both aggradational and degradational.

The Measurement and Description of Slopes

Accurately determined slope profiles may be measured in the field, or less satisfactorily, from high-quality detailed maps. The investigator measures the slope angle (inclination from the horizontal) at its maximum steepness along a profile line running from the hillcrest to an adjacent stream channel. In the field, a **clinometer** is used to make repeated measurements of the angle along the profile line over a series of regular measured distances. The profile thereby recorded and drawn on a graph can then be described in several ways.

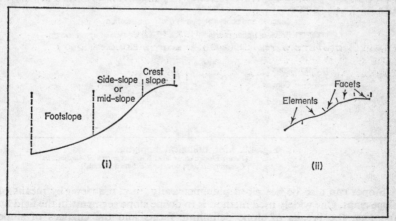

Fig. 4.1. Slope description: (i) parts of a slope; (ii) segments of a slope.

29

Fig. 4.1 shows two of these: in the left part of the diagram (i) is a simple description of the main parts of a slope; to the right (ii), the slope is described in terms of slope **segments**. A slope **facet** is a straight segment of a slope profile; a slope **element** is a smooth convex or concave segment. The terms facet and element can also be applied to areas of slope in the field.

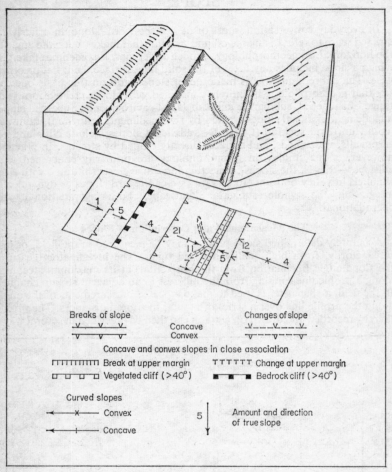

Fig. 4.2. Morphological mapping.
(From Doornkamp and King, *Numerical Analysis in Geomorphology*, courtesy Edward Arnold)

Slopes can also be described planimetrically (in plane view) by means of **slope maps**. One widely used method is to define slope segments in the field by mapping the breaks of slope delimiting facets and the changes of slope between the elements. Fig. 4.2 demonstrates a model example of this.

Approaches to Slope Studies

Geomorphologists have spent a lot of time arguing about slopes. There are several reasons why slope studies have proved difficult. First, despite the need for all slope forms to be measured with precise field survey methods, many have been simply estimated by eye, or drawn from maps with large contour intervals. Since many early writings on slopes were based on visual impressions rather than accurate information, it is perhaps not surprising that certain misconceptions about slope form have persisted unchallenged for years. One false impression is that the British landscape is made up of concavo-convex slopes, but a few accurate measurements will prove that rectilinear facets also abound. Second, slope processes are complex, and vary in their importance on different parts of a slope. It is very difficult to unravel cause and effect, and the relation between form and process is often very much a chicken and egg problem. Does the slope angle control the process, or the process control the slope angle? Third, slopes have been at the centre of several broad schemes of landscape evolution proposed by various geomorphologists, and this has meant that some rather doctrinaire attitudes often prevail over slopes. The best known of these schemes is that proposed by W. M. Davis and called the **Geographical Cycle**. The wider implications of this cycle will be discussed in Chapter Nine.

In the last twenty years or so there has been something of a revolution in slope studies. Formerly, there had been a tendency to assume that present-day processes working on slopes could be largely ignored, as they were either too slow or too catastrophic to measure. But improved techniques have rendered even the slowest movements measurable. Thus many slope studies have now become field orientated and at least partly quantitative.

It is now generally accepted that slopes cover a wide range of adjustment between form and process; that is, some slopes are in intimate equilibrium with present process, whereas others are not at all. We can therefore bear in mind two approaches to slopes: an **historical** approach in which slopes are regarded as a reflection of what has happened in the past, and a **dynamic** approach which regards slopes as being in adjustment with reasonably contemporaneous processes. Both approaches are valid and indeed necessary to the proper understanding of slopes.

Slopes as Open Systems

Since slopes are rather complicated in their mechanisms, one useful way of studying them is to regard them as open systems. This helps us to identify some of the forces and processes at work. It also allows us to regard slopes as continually striving to approach a condition of equilibrium. Fig. 4.3 represents one attempt to view a hillslope as a dynamic open system. The slope in this case is regarded as a solid figure, in which the watershed and river channel mark the boundaries at the top and bottom of the slope, and the ground surface and solid bedrock beneath the soil layer mark the upper and lower boundaries on the slope itself. Across these boundaries, movements of energy and matter take place, causing the slope processes and the adjustment of form to them.

The main inputs of energy and mass into the system include the potential matter and energy of the available relief, matter in the form of precipitation,

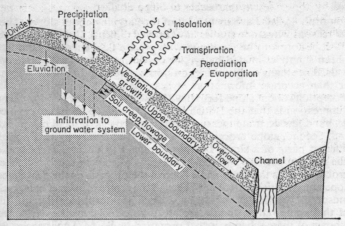

Fig. 4.3. A hillslope considered as an open system.

and energy from direct solar radiation and thermal radiation from the air. Some of this energy is lost back into the atmosphere by conduction, evaporation and outgoing radiation. Material passes through the system by infiltration, carrying with it ions and particles of clay and silt (the process of **eluviation**) and by overland run-off, which may initiate gully erosion. Material is also transported by mass movement, including slope creep and soil flowage. There is inevitably a tendency for downslope movement because gravity exerts a bias on the individual processes at work. The material is lost to the system at its lower end into the adjacent stream channel. The activity in the stream adjusts itself to the material supplied to it from the slope system. This removal of solids and solution from the system contributes to changes in the shape or position of the slope.

Mass Movements

The wide variety of agencies which erode and move material in the slope system are normally grouped into two main mechanisms: (*a*) mass movements, involving the removal of large portions of the hillside under gravity; and (*b*) the action of water on slopes. Despite this division, we may note that both groups are part of a general continuum of processes: the difference, for example, between a muddy waterflow on a slope and a runny mudflow is necessarily arbitrary. This continuum has been used by some authors to classify mass movements, at one extreme embracing types in which there is a great dominance of water over debris, as in mudflows, and at the other, rockfalls, in which there is very little water involved.

Mass movements include the mechanisms both of **flowage**, in which the velocity of the flow is greatest at the surface, and **sliding**, where the velocity at the base is similar to that at the top. In either case, the downslope movement of materials is a response to the application of **shearing stresses**, caused by gravity and the weight of the material and soil water. These forces increase with increasing angle and height of slope. **Resistance** is provided to these stresses by the cohesive properties of the soil particles and their internal friction. This, in turn, is dependent on the pressure exerted by the soil water occupying the pore spaces of the soil. The role of water in mass movements is thus very large. The introduction of high intergranular water pressures can effectively reduce soil strength, and additional water increases the weight of slope materials. Hence, we may expect many landslides to occur after heavy rain.

The large variety of mass movements makes it impossible to build up an entirely satisfactory classification. Here, a simple division into creep, flows, slides and slips, and falls is used.

Creep and Other Types of Slow Flow

The slow downhill movement of debris and soil is described as **creep**. Fig. 4.4 gives some indication of the variety of phenomena that may give

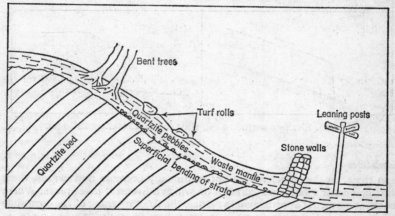

Fig. 4.4. Types of evidence indicating soil creep.

evidence that creep has been taking place: leaning fence posts, trees and poles; the accumulation of earth on the upslope side of stone walls; the bending over a bedrock strata near the surface. The cause of creep almost always lies in the combination of processes, including rainsplash impact, frost action, cracking caused by drying, and the activity of animals and plant roots. To these must also be added the effect of ploughing, which causes appreciable downslope movement of material. Other types of creep which may be recognised are talus creep, rock creep and rock glacier creep; these are all essentially movements of coarse debris in various environments. **Solifluction** is the slow flowage of soil taking place in periglacial regions, and will be considered in more detail in Chapter Six.

Rapid Flows

These depend on there being sufficient water to saturate comprehensively the soil mass. **Mudflows** and **earthflows** (Fig. 4.5) have bowl-shaped source areas leading to a long chute through which the slope material rapidly passes before spreading out in a series of lobes in the depositional area. Contrary to many earlier ideas, it has been observed that many flows are bounded by well-defined shearplanes, between which plug flow may take place. The movement may be catastrophic. Allied to mudflows are **bog bursts**, sudden outflows of water and plant debris which sometimes occur in saturated peat bogs.

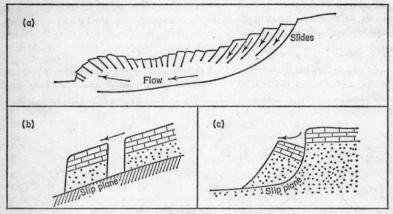

Fig. 4.5. Idealised earthflow (a), slide (b), and rotational slip (c).

Landslides

In these, as the velocity does not continually decrease downwards, there must be one or more shear surfaces on which movement takes place. Where the shear surface is approximately planar, the strict meaning of the term **slide** is appropriate. However, another common type of landslide takes place on arcuate shear planes, and these are called **rotational slips** (see Fig. 4.5). The latter type of landslide is common on the south coast of England in eastern Devon and at Folkestone Warren in Kent.

Most landslides usually occur fairly rapidly, often after excess groundwater following heavy rain has reduced soil strength. A weak stratum, perhaps a bed of clay, often provides the zone in which shearing eventually takes place. In homogeneous rocks, landslides may sometimes be started by the removal of support from the front of a slope, perhaps by sea erosion or artificial excavation.

Rock and Debris Falls

These are liable to occur wherever steep slopes are maintained by erosion, for example on sea cliffs or on glacial headwalls; or where man has artificially created precipitous slopes in road and railway construction. The actual fall is set off by some trigger mechanism. A Scandinavian geomorphologist, A. Rapp, has listed a number of possibilities: frost bursting, heavy rain, chemical weathering, thermal changes and wind action.

Run-off on Slopes

The action of water is one of the most ubiquitous processes on hillslopes. Basically, two factors determine whether there will be any surface run-off and what type of flow will take place: the rate of rainfall (note, not the *total* amount); and the infiltration capacity of the soil. In addition, water may be introduced to the slope by groundwater springs.

Raindrop Impact

The impact of raindrops can cause a surprising amount of disturbance to the surface soil and we can speak in terms of **splash erosion**. On a level surface rainsplash has been observed to move particles 4 mm in diameter a distance of 200 mm. The effect of impact on a slope is to move stones in all directions, but with a net result downslope. Large stones can be moved by rainsplash by the process of undermining. Raindrop impact is not usually effective when the rate of rainfall is light; surface binding of colloids may make the soil sticky and cohesive. Vegetation has an important role to play in determining the amount of exposed bare surface liable to splash erosion; it is perhaps not surprising that splash erosion is probably most important in semi-arid regions.

Subsurface Flow

Water penetrating into the soil may either join groundwater at depth, in which case it has little immediate effect on erosion, or it may flow laterally, in which case it is usually referred to as **throughflow**. Throughflow velocities are usually fairly low, and therefore not necessarily a potent agent of erosion. However, concentrations of throughflow (**piping**) may lead to the heads of surface gullies, and help to extend those gullies by eroding subsurface material.

Overland Flow

Run-off that flows down the slopes of the land in more or less broadly distributed films, sheets or rills is referred to as overland flow—as distinct from channel flow, in which the water occupies a distinct trough confined by lateral banks. In temperate regions, the soil is normally capable of absorbing a considerable amount of rainfall by infiltration. If the rainfall is not intense, overland flow may not occur at all, or at least, not for some time. With heavier rainfall, soil passages may become sealed or obstructed causing overland flow to commence. Initially, **surface detention** will hold some of the surface water in small puddles, but these will gradually fill up and overflow. As the water moves down the hillside, we can expect the volume of water to increase proportionally to the length of the slope. Thus either the depth of the water increases or the flow velocity increases, and erosion may be initiated. At the foot of the hillslope, the overland flow passes into a stream channel or lake.

Overland flow is potentially an important factor in the erosion of hillsides. Under stable natural conditions, the erosion rate in a humid climate is slow enough to permit the soil cover to replenish itself, allowing a vegetation cover to be maintained. But disturbance of the natural equilibrium by man, frequently through the removal of the vegetation cover, can lead to a state of **accelerated erosion**. Destruction of the vegetation greatly increases the likelihood of splash erosion, and drastically reduces the resistance of the ground surface to erosion by overland flow. We find that the eroding capacity of overland flow, besides being directly proportional to the rate of precipitation and length of slope, is also inversely proportional to the infiltration capacity of the soil and the resistance of the surface.

The results of accelerated soil erosion are well-known. Large-scale sheet removal of soil layers and deep gullying have affected many parts of the world, not least the Mid-west of the United States in the 1920s and 1930s.

An interesting model of slope development, relating overland flow to erosion, was proposed in 1945 by R. E. Horton, and has become known as the **hydraulic slope theory**. The broad principles are illustrated in Fig. 4.6. Rain falls uniformly over the hillside, and the amount of run-off increases at a uniform rate downslope. Three sectors on the slope are recognised: in the upper part of the slope, there is a belt of no erosion, where the overland flow

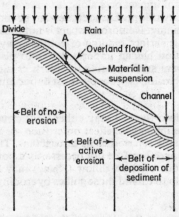

Fig. 4.6. Erosion zones related to overland flow.
(Courtesy The Geological Society of America)

is not of sufficient force to overcome the resistance of the soil particles. At point A, however, this is exceeded, erosion sets in, and material is carried in suspension. This is denoted as the belt of active erosion. Towards the foot of the profile, the slope gradient decreases and some of the suspended matter is dropped as colluvium.

This theory has been tested and proves useful as a reasonable guide as to what happens on such a profile. But it does not explain how the profile developed in the first place or what its future evolution will be. Finally, most slopes seemed to be formed under the combined action of mass movement and run-off, rather than one or the other.

Controls on Slope Processes

Simple correlations between slope form and process rarely exist. The factors which control the type and rate of processes are often cited in relation to the slope form itself; for instance, one speaks of the relationship between climate and slope form. But we should always remember that the control operates through the processes. Some of the factors which control hillslope processes are: the nature of soil materials and bedrock which comprise the slope; the prevailing climatic conditions; the vegetation cover, itself largely dependent on climate; downslope factors operating at the foot of the slope, e.g. stream channel activity; and available relief. The more important of these factors will be briefly reviewed here.

Lithology and Structure

A great many observations have been made on the subject of the geological control of slope form, but as usual the problem is that rocks and structures act in combination with other factors to influence process and produce complex landscapes. Rock type influences the slope angle both directly and through its control on the nature of superficial deposits (or regolith characteristics).

Resistant rocks normally produce slopes that are steeper than those on weak rock types. The actual mechanics by which this achieved is related to such factors as the cohesion of the rock and the relation of joints and other weaknesses to the slope angle. However, it is worth noting that so-called weak rock types which are highly cohesive, such as some clays, may also form steep slopes. We must not assume that a particular rock type always produces a particular slope profile. This may well apply locally or regionally, but on a global scale, rocks may behave differently because of changes in their weathering resistance.

Layered sedimentary rocks of alternating resistance create marked relief effects (Fig. 4.7). Where dips are almost horizontal, isolated scarped ringed hills, or **mesas**, may be produced. Where the rocks are gently dipping, a **cuesta** form is typically in evidence, consisting of a steep **escarpment** and a gentle **dipslope**. The scarplands of south-east England, South Wales and the Wenlock Edge area are all good examples of cuestas. As the dip of the rocks becomes steeper, slopes become increasingly regular until a symmetrical **hogsback** results.

Escarpments themselves have been relatively well studied. S. A. Schumm

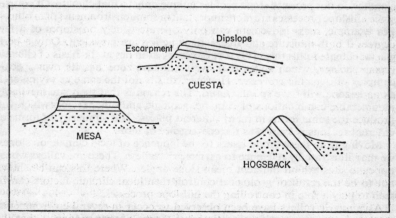

Fig. 4.7. Landforms on layered sedimentary rocks.

and R. J. Chorley, working in the Colorado Plateau, concluded that the actual form of the scarp depends on four variables. First, rock resistance, as controlled by the strength of the cementation of individual particles and by the porosity of the rock; second, by the orientation and spacing of joints and bedding planes; third, by the way in which the rock is dipping, i.e. away from or into the scarp; and fourth, by the thickness of the cap-rock at the top of the scarp—once a cap-rock disappears, scarp slopes may rapidly degrade. Of these points, the infiltration capacity of rocks appears to be one of the most important factors controlling slopes. Highly permeable limestones often have very steep slopes; chalk slopes are frequently convex. On impermeable rocks on the other hand, such as clays, the high surface run-off produces slopes that are often concave.

Slopes and Climate

Since climatic parameters such as temperature and precipitation amount are bound to influence the rates of weathering and run-off processes, then it could be argued that climate is bound to influence slope form. But there has been much disagreement over this. On the one hand, some geomorphologists, such as the South African, Lester King, have forcibly argued that slopes develop similarly in virtually all climates, with micro-climate accounting only for minor aspects of slope morphology. On the other hand, other geomorphologists seem to believe that slope forms do vary with climate. Broadly expressed, this view can be stated in the following terms. Slopes in humid climates are smooth, relatively gentle, mantled with soil and vegetation, and consist of convex-concave profiles which decline with time. Slopes in arid climates are rough, relatively steep, barren of soil and vegetation, and consist mainly of straight slopes which retreat parallel to themselves to produce pediments (see Chapter Seven).

Detailed field studies have not always borne out these broad climatic precepts. It has been amply demonstrated in America and England that valley side slopes are often straight. The general conclusion we have to come to is that all kinds of hill forms are found in all kinds of climate, i.e. specific climates do not produce slope profiles unique only to that climate. Of course, some hillslope processes are more important in some environments than others. For example, mass movement can only operate widely on slopes of a few degrees if both moisture and freeze–thaw action are prevalent. Only a periglacial climate fulfils this. This sort of observation lies at the basis of **climatic geomorphology**, which states that every climatic zone has its own specific complex of dynamic *processes*. However, this is not the same as saying each climatic zone will have specific *forms*. The reason is that there are inevitably innumerable combinations of climatic, geologic and other factors which will produce the same forms in many different places. This similarity of form for different reasons is known as the **convergence of form**.

Moving from these broad issues to the influence of local climate on slopes, we may illustrate by reference to **asymmetric valleys**. These are valleys whose opposite sites exhibit different mean slope angles. Where this can be shown not to be the result of geological control, then local climatic factors can be seen to play a role in controlling the hillslope processes.

Asymmetric valleys have been observed to occur in several environments, so there is no one specific climatic cause. Some observers suggest that the more exposed slope, facing the prevailing wind or having the most sunshine, is the steeper, because it is being actively eroded. Alternatively, the eroded colluvium accumulating at the base of a slope may force the valley stream to undercut the opposite side, contributing to its steepness. Many asymmetric valleys may originally have been formed in the last Ice Age. The steeper south-west facing slopes in the Chilterns, England, have been explained as relict features of former periglacial conditions. These slopes were more prone to solifluction and frost weathering because they received more insolation and underwent more freeze–thaw cycles. North-east-facing slopes, being more shaded, were frozen through the day and therefore less active.

Slopes, Relief, and the Drainage Basin

The amount of available relief—the difference between the highest and lowest points in the landscape—and the degree of dissection are both bound to affect slope form. Slope angles tend to be greatest in areas of high relief and where dissection is rapid, whereas in stable, low-relief areas, slopes are usually less steep and more concavo-convex in profile. The connection lies in the activity of the valley streams. It is impossible to consider slopes and streams as completely separate entities, as they are parts of the same open system, the drainage basin. The channel at the foot of the slope will exert a considerable influence on the hillslope form. On the one hand, the stream will tend to adjust its gradient so it can transport away the amount of debris the slope supplies to it. On the other, the slope form will adjust so that it supplies the load which the stream will be able to carry. In other words, they interact. When the stream-slope system is mutually adjusted, i.e. in equilibrium, then the drainage basin can be considered to be in a steady state.

The mutual interaction of slopes and streams therefore means that if there were to be some change in the stream channel gradient, perhaps because of tectonic activity, or because of a variation in water discharge consequent on climatic change, then this will materially affect the slope form. Hence, rejuvenation of the stream will lead to further erosion of the valley sides.

Slope Evolution

Much of the discussion about the development of slopes through time has been dominated by the views put forward by W. M. Davis and W. Penck. Davis postulated that in a 'normal' cycle of erosion, slope profiles become progressively smoother and steadily decline in angle. Penck considered that slope replacement and parallel retreat took place on some parts of a slope (Fig. 4.8). Other workers such as A. Wood and Lester King have also accepted parallel retreat and slope replacement as the primary modes of change.

Numerous local studies have attempted to evaluate the relative importance of **parallel retreat** versus **slope decline**. One important fact that seems to emerge is that in rapidly eroding areas, as in badlands, slopes retreat parallel to themselves providing the eroded material is removed from the base. If, however, the slope debris is not transported away from the slope base, the slope angle declines. Thus one may have parallel retreat and slope decline in the same environment depending on whether or not a stream, or alternatively the sea, is undercutting at the base of the slope.

One well-documented example demonstrating this is a study by R. A. G. Savigear on slopes in Carmarthen Bay, South Wales. He measured a number

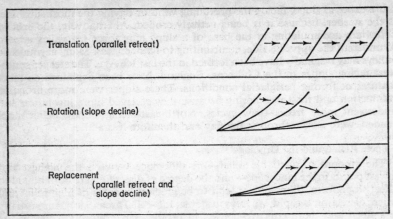

Translation (parallel retreat)

Rotation (slope decline)

Replacement
(parallel retreat and
slope decline)

Fig. 4.8. Types of slope change.

of slope profiles on an Old Red Sandstone cliff, which has been gradually cut off from the sea by the development of a spit at its foot. On the most recently abandoned part of the cliff, the slope profiles are dominated by a long recti-linear section with a mean angle of 32° (profile A, Fig. 4.9). The rectilinear section appears to be a good example of a debris-controlled slope, the debris being removed by gravity fall over the former marine cliff. As long as this condition of unimpeded removal continues, then the slope, although eroding, will maintain its characteristic angle of 32° by parallel retreat. However, on the long-abandoned part of the old cliff (profile B) debris is no longer being removed at the same rate as formerly and a basal concavity has developed at the lower end of the slope. In this situation of impeded removal, the rectilinear section has an angle of 28°; in other words, slope decline has taken place.

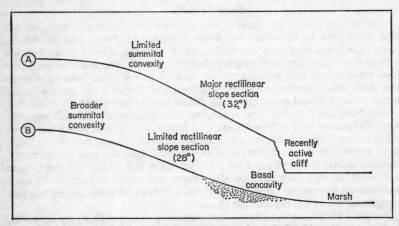

Limited
summital
convexity

A

Major rectilinear
slope section
(32°)

Broader
summital
convexity

B

Limited rectilinear
slope section
(28°)

Recently
active
cliff

Basal
concavity

Marsh

Fig. 4.9. Cliff profiles from South Wales (after Savigear).

The issue of the dominant mode of slope evolution is a central issue in the study of the total evolution of landscapes, and this will be considered in Chapter Nine. We may conclude by saying that as far as slopes are concerned, it is clear that no single theory of hillslope evolution can be demonstrated. Slope forms depend on a number of interrelated variables such as lithology, climate, vegetation and relief, which operate with varying intensities in different environments. Different processes can produce similar forms, and hence processes can never be inferred from form alone.

Suggested Further Reading

Brunsden, D. (ed.), *Slopes Form and Process*, Institute of British Geographers Special Publication, London, 1970.

Carson, M. A., and Kirkby, M. J., *Hillslope Form and Process*, Cambridge University Press, London and Cambridge, 1972.

Fairbridge, R. W. (ed.), *Encyclopedia of Geomorphology* (Slopes), Reinhold, New York, 1968.

Young, A., *Slopes*, Oliver & Boyd, Edinburgh, 1972.

CHAPTER FIVE

RIVERS AND DRAINAGE BASINS

A river or stream is a body of water flowing in a channel. Rivers have always been of supreme importance to man, providing focal points for habitation, water for cultivation and avenues of travel, water power and recreation. They are even more fundamental in the natural world. The action of running water is the most ubiquitous landscape-forming agency. This chapter will consider not only the landforms created directly by streams, but will also explore the relationship between streams and other components of drainage basins.

Hydrology

All rivers ultimately receive their water from precipitation. However, the relationship between precipitation and river discharge is not always a simple one; only a small amount of the water normally reaches the stream channel. The behaviour of water in this respect is best expressed as part of the **hydrological cycle**. Water falls on the Earth as some kind of precipitation— rain, snow, hail or sleet. Part may evaporate back into the air or be transpired by plants; part may be temporarily retained on the surface in lakes or glaciers; part may run off immediately into streams; part may be retained in the plants or in the soil; and a part may sink into the ground to become **groundwater** and reach the stream at a later time as groundwater flow. Through the rivers, the water reaches the sea, to be evaporated again.

In studying the hydrology of rivers specifically, the framework of the **basin hydrological cycle** is used, in which the drainage basin is taken as the focus of study rather than the global hydrological cycle. We can then view the basin cycle as an open system with inputs of precipitation (p), being regulated through the system by various means of storage, leading to outputs of basin channel run-off (q), evapotranspiration (e), and outflow of groundwater (b). Fig. 5.1 represents a detailed attempt to systematise the relationships between various types of water in the basin. A careful perusal of the flowchart in the lower part of the figure will explain the symbols used in the block diagram of the basin. The groundwater zone is normally divided into a **zone of saturation**, where the underground water fills all the spaces in the rock, and a **zone of aeration** above it, in which the water does not fully saturate the pores. The **water table** marks the change from one zone to the other.

We can see now that much of the flow of a river is not from immediate run-off; groundwater storage is important in regulating the flow of a river and supplying water to it in between periods of rain. The **regimen** or habit of a stream thus depends on three main factors: the amount and intensity of rainfall, itself governed by various climatic factors; the infiltration capacity of the soil and rock over which it is flowing; and the morphological properties of the drainage basin. The nature of the vegetation cover also has an important role to play.

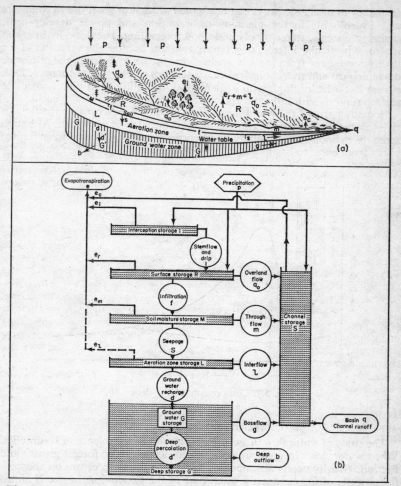

Fig. 5.1. The basin hydrological cycle: (a) block diagram of the basin; (b) flow diagram of water movement.

Stream flow, as measured by the hydrologist, is recorded in the form of a **hydrograph**, which shows the variation of discharge with time. Fig. 5.2 shows the principal components of a hydrograph. It is usually possible to distinguish the level of **base flow**, resulting from run-off supplied by groundwater, from that of **flood flow**, which produces a series of sawtooth-shaped fluctuations in the hydrograph. One such peak is shown in the diagram. The peak of discharge is usually obtained sometime after the most intense rain has ceased (basin lag).

From such hydrographs, much information about the magnitude and frequency of discharge peaks can be gleaned which is of great use to engineers and hydrologists in the planning of irrigation and power development, drainage systems, water supply and flood forecasting. The flood hydrograph is also of particular interest to geomorphologists because the variation of the hydrograph shape from basin to basin shows the dependence of the discharge on geological and morphological characteristics of a drainage basin. A **flash stream** has a hydrograph with a sharp peak, resulting from high immediate surface run-off, with little absorption and storage of water in the basin.

In many parts of the world, spring is a time of floods because the flow of rivers is augmented not only by spring rains but also by snowmelt. Although some rivers flood quite regularly, the **recurrence interval** (time between floods) varies from basin to basin, and varies with the magnitude of the flood.

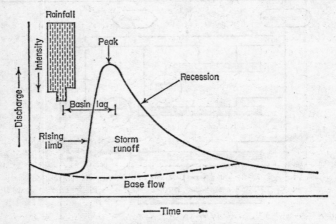

Fig. 5.2. Components of a hydrograph.

Water Flow in Rivers

The study of water flow in a channel is known as the science of Hydraulics. When water flows in a stream it is subject to two basic forces, gravity and friction. **Gravity** exerts an impelling force which puts pressure on the confining walls of the channel; a small part of the gravitational force is aimed downstream causing flow. Opposing the downstream flow is the force of **frictional resistance** between the water and the bed of the channel.

The water flow is not steady and uniform, but in all but the most sluggish streams, is affected by **turbulence**. This takes the form of a variety of chaotic movements and eddy systems. Turbulence in streams is extremely important because it creates upward motion in the flow which lifts and supports fine particles of sediment. If **laminar flow** were to prevail in streams, whereby the water flowed in parallel horizontal layers, then any sediment would remain on the bed. But laminar flow is rare in natural streams, although it is common in groundwater movement.

The effect of friction ensures that the velocity distribution of water flow in a river is not even; water closest to the banks moves only slowly, whereas that near the centre moves fastest. The highest velocity is usually located in mid-stream about one-third of the distance down from the surface to the bed. Fig. 5.3a shows a typical velocity distribution in a symmetrical channel. The precise shape of the channel has a marked effect on velocity: in an asymmetrical channel, the zone of maximum velocity shifts away from the centre towards the deeper side. This in turn also causes the zone of maximum turbulence to become lower on the deeper side (Fig. 5.3b). In this way, significant erosive effects are brought about by change in channel shape.

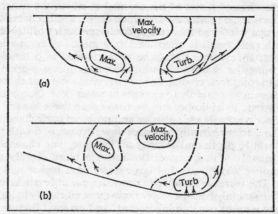

Fig. 5.3. Zones of maximum velocity and turbulence in (a) symmetrical and (b) asymmetrical stream channels.

Because the velocity varies so much across the stream, a single figure expressing the mean velocity, V, is usually calculated for the entire cross-section. This figure will vary principally with the gradient of the stream, the volume of water and the roughness of the bed, as well as with the width, depth and shape of the channel. There are various formulae available to quantify the velocity. Another useful measure of stream flow is the discharge Q, defined as the volume of water passing through a given section of stream in a given unit of time, in practice usually stated in cubic metres per second. Discharge may be obtained from the formula $Q = AV$, where A is the cross-sectional area.

Stream Energy

Any river's ability to do work—that is, to erode and transport material—depends on its energy. Potential energy is provided by the weight and the elevation of the water. This is converted by gravity into downflow and hence into kinetic energy. However, something like 95 per cent of this energy is lost owing to friction within the water and between the water and the banks. The roughness of the channel has a marked effect here: a rough channel, perhaps strewn with boulders, creates considerable eddying and loss of energy; a smooth channel on the other hand decreases the frictional loss, making more energy available for work. The efficiency of the cross-sectional form of the channel is often measured by a quantity known as the **hydraulic radius**, which is defined as the ratio between the cross-sectional area and the length of the wetted perimeter. The higher the ratio the more efficient the stream and the smaller the loss due to friction. The ideal form of a channel for the discharge of water would be one which was semicircular in cross-section, but of course very few natural channels possess this.

Channel Shape

A close relationship exists between the velocity and discharge of water in a river and the characteristics of the channel in which the water is flowing. These characteristics, or parameters, include depth, width, channel roughness, and are collectively referred to as the **hydraulic geometry** of the channel. The width to depth ratio (w/d) is often used as a means of comparing different channel shapes. In essence, the shape of a channel is determined by the materials forming the channel sides and the river forces working on them. Channels cut in solid rock change only slowly with time, but those composed of river alluvium adjust rapidly to changes in water flow. Normally, channels in silt and clay tend to be deeper and narrower than those in sand and gravel, because the finer materials are cohesive and promote bank stability.

Discharge in a river normally increases downstream, and with this, we find that channel width, depth and velocity all increase, but channel roughness decreases. Channel width increases downstream more rapidly than depth, which put another way means that large rivers have higher w/d ratios than smaller ones. The increase in velocity downstream, at levels below bankfull discharge, occurs mainly because water flows more efficiently in large channels and less energy is spent overcoming internal and external friction. Hence the carrying capacity of the stream is increased and a lower gradient is required to transport the load. River gradients also tend to become less steep downstream as the load gradually abrades itself and becomes of lesser calibre and easier to transport. On the occasions when the river is at bankfull discharge, velocity appears to be more or less constant downstream.

Transport, Erosion and Deposition

Transport

A stream carries material in three ways: the **dissolved load** is composed of soluble materials transported invisibly in the form of chemical ions; the **suspended load** consists of clay and silt held in the water by the upward elements in the turbulent motion; larger materials move as **bed load** close to the channel floor by saltation, rolling, and sliding. The percentage contributions to the total load vary widely with the nature of the river and the season at which observations are taken. In North America, the suspended and dissolved loads are roughly comparable on average, but the amount of bed load is more difficult to assess. Semi-arid rivers such as the Missouri and Colorado have very high suspended loads because of the high rates of sediment supply to the channel by overland flow.

The load carried by a stream increases with increased discharge and velocity. We refer to stream **capacity** to denote the largest amount of debris a stream can transport, and to stream **competence** in relation to the diameter of the largest particle that can be moved. The lowest velocity at which grains of a given size move is said to be the **critical erosion velocity**. The relationships between size and velocity are shown in Fig. 5.4; an interesting point here is that sand is more easily moved than silt or clay. This is because the fine-grain particles tend to be more cohesive. High velocities are required to move gravel. A wide area, rather than a line, is used in the graph to define erosion velocity, because the value of erosion velocity varies with the depth and temperature of the water and with the density of the grain.

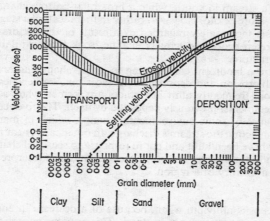

Fig. 5.4. Erosion and deposition in relation to stream velocity.
(From B. W. Sparks, *Geomorphology*, Longman)

Erosion

Streams erode in various ways, depending on the nature of the transported material, if any, in the river. The force of flowing water alone is termed **hydraulic action**, and this can exert a dragging effect upon the bed, eroding poorly consolidated materials such as sand, silt and clay. Two types of hydraulic action are sometimes distinguished: **evorsion**, the direct impact of water; and **cavitation**, a pressure effect under very high velocity flow. Erosion by solution is termed **corrosion**. However, the principal means of erosion is **corrasion** or abrasion. This is the mechanical impact produced by the debris carried by the stream. It not only erodes the bed of the stream, but also the entrained particles themselves, grinding them into smooth-rounded shapes of different sizes. One result of stream abrasion is the **pothole**, a cylindrical hole carved into the hard bedrock of a stream. Other features produced by abrasion include plunge pools, chutes and troughs on the stream bed.

Erosion takes place where the stream has an excess of energy, but excess energy does not always result in erosion; much depends on the resistance of the bed over which the water is flowing. Erosion by running water is fundamental in initiating and then developing channels. Once overland flow on hillslopes has become concentrated into gullies, these can develop by headward and lateral erosion. **Headward erosion** results from undercutting at the base of the soil or rock or of a vegetation mat. The undercutting is often caused by percolation underground, which removes fine material and develops subsurface pipes and tunnels, which become exposed as open gullies as they widen and cave in. **Channel widening** takes place by erosion against the sidewalls, particularly on the outside of river bends when the stream is in flood. Channel widening along with weathering and hillslope processes contributes to the overall widening of the valley. **Valley shape** has traditionally been ascribed to stages of an erosion cycle. A V-shape valley has been regarded as the result of a stream in youth, while a broad flat-bottomed valley has been taken to be mature or old. However, rather than indicating a stage of erosion, valley shape is more safely regarded as a result of the factors that control slope and stream processes, namely climate, rock type, available relief and geological structure.

Terraces are a landform contributing to valley shape and are usually the result of both erosion and deposition. Terraces may be benches cut in solid rock, but more frequently alluvial terraces are formed when a river erodes flood-plain sediments, previously deposited by itself. The river cuts into these deposits because of some environmental change, which in many cases is a climatic one affecting the stream's discharge. In other cases, near river mouths, terraces may have been built and cut in response to sea-level changes. Terrace sediments and morphology are often used as guides in interpreting the geomorphological history of a region.

Deposition

A river deposits alluvium when, because of a decrease in energy, it is no longer competent to transport its load. This usually occurs because of a reduction in the gradient of the stream channel, but may also result from an increase in the calibre of the load, perhaps brought in by a tributary into the main stream, or by conditions of accelerated erosion upstream. The first debris to be deposited will be the largest calibre, succeeded downstream by finer material, while the very finest material may continue to be transported even although the river energy has been reduced. This sequence of sedimentation is found in many of the depositional forms created by rivers.

A **flood-plain** is the most common depositional feature created by all sizes of river, be they very large or just small brooks. The alluvium in a flood-plain is composed of several kinds of deposit. Principally, the flood-plain is built up laterally by channel deposits in a coalescing series of **bars** composed of sand and gravel scoured from the outside of meander bends upstream; and it is also built up vertically by aggradation of overbank deposits during flooding. Floods deposit coarse material as **levees** near the channel, and finer silt and clay over the rest of the flood-plain surface, covering bars previously formed by the river. Flood-plains are characterised by a distinct collection of landforms, such as **oxbow lakes**, sloughs, ridge and swale topography and **backswamps**, some of which are shown in Fig. 5.5. More will be said of the role of meanders in building up the flood-plain later in the chapter.

A landform often built by streams carrying a large load of coarse debris is the **alluvial fan**. Where such a stream flows out of a hilly area on to the gentler slope of a plain, its gradient and velocity are drastically reduced, forcing the stream to rapidly aggrade. There may also be a loss of volume by percolation. The deposition causes the stream to divide and change its course frequently,

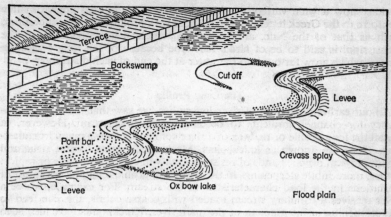

Fig. 5.5. Flood plain features.

building up a fan in a cone-shaped form, with a fixed apex where the stream emerged from the hilly area, and a downward slope in all directions from the apex (Fig. 5.6a). There will be a gradation in sediment size down the fan from coarse to fine. Sometimes several individual fans merge in a line along a mountain front to form a **bahada**, which are seen mainly in arid regions.

Where streams flow into standing water, a **delta** may form, deposition being caused by the rapid reduction in stream velocity as the stream current pushes out into the lake or sea. A regular succession of deposits forms (Fig. 5.6b) in which the finest particles are carried furthest, creating **bottom-set** beds; coarser material is deposited as a series of steep-angled wedges (**fore-set beds**) as the delta progrades into the water, and the coarsest material is carried in the river channel and dumped on the surface of the delta as **top-set beds**.

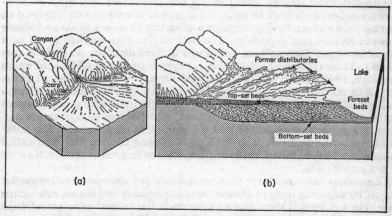

Fig. 5.6. (a) An alluvial fan. (b) A delta in vertical section.

Deltas show a wide variety of shape in plan view, but all approximately conform to the Greek letter delta (Δ). Those with broadly curving shorelines, such as that of the Nile, are known as **arcuate deltas**; the delta of the Mississippi is said to be of **bird's foot** type because of the long projecting fingers which grow far out into the water at the end of each distributary.

The Long Profile

In our earlier discussion of river hydraulics, we saw that there was a tendency for channel gradients to become flatter downstream. However, in detail the long profile or **thalweg** of a river commonly shows many irregularities. These are known as **knickpoints**. Many knickpoints have a structural or lithological origin: bands of resistant rock will create waterfalls or rapids. Other, more subtle steepenings in the river long profile might be the result of variations in the load characteristics in the stream. For example, where a river receives a tributary stream loaded with coarse debris, this can lead to the steepening of the gradient of the main river. Knickpoints have also been regarded as a result of changes in sea-level, steepening the thalweg in the lower part of the stream.

Grade and Steady State

In time, irregularities such as rapids and waterfalls will be eroded away by the river. Observations show that irregularities on alluvial channels disappear very rapidly. Rivers tend to attain a condition of equilibrium which is called **grade**. The term was first introduced by G. K. Gilbert, an American geologist, but W. M. Davis enlarged on the theme to fit in with his theory of cyclic development (see Chapter Nine). Davis regarded grade as a condition of balance between erosion and deposition, brought about by the ability of the river to adjust its capacity to the amount of work being done. He envisaged that the adjustments made by the stream in reaching and maintaining grade were principally adjustments in channel gradient. He linked the concept of grade with the attainment of a smooth concave upward profile.

Although most streams tend towards this general outline, most geomorphologists find it difficult to recognise a graded stream as Davis defined it, because even streams having irregular profiles can be shown to have a balance between erosion and deposition. The best way to regard grade is not as a two-dimensional phenomenon, but as one in three dimensions; grade is now taken to be as Gilbert first defined it—namely a condition of balance where the slope, width, depth and other channel characteristics are adjusted to the prevailing volume of water and the load it is carrying. This cannot be a short-term equilibrium, because erosion and deposition are taking place almost continuously, but viewed over a long term, such changes tend to cancel out. Thus, for example, a gravel bar that is deposited in the channel at low water will be removed in flood. This kind of oscillating balance has been called **dynamic equilibrium**, but using the nomenclature of an open system, such a river is in a **steady state**.

This state is self-regulatory; the river reacts to any change of environmental factors by adjusting itself to absorb the change and re-establish a new steady state. For example, if there was an increase in the volume of water supplied to a river which was already in a steady state, then this would result in a change in channel characteristics. These would adjust to carry the new volume: erosion might change the depth, width and channel roughness.

In summary, then, the term 'steady state' can be applied to rivers which have reached a state of self-regulation and maintain stable channel characteristics. They do so by adjusting their long profile, cross-sectional shape and channel roughness.

Channel Patterns

If we look at any river on a map, we observe that it is either straight, crooked, meandering or braided (separating and rejoining). Straight channels are rare; most natural streams wander. In their study of channel patterns, geomorphologists have paid particular attention to meandering and braiding, which provide good examples of the close relationship between water flow, processes and the landforms produced.

Meanders

Meanders are sinuous bends of a highly characteristic form. The degree of meandering may be described on a scale of sinuosity, which relates channel distance to axial distance (Fig. 5.7a). In addition, meanders can be described

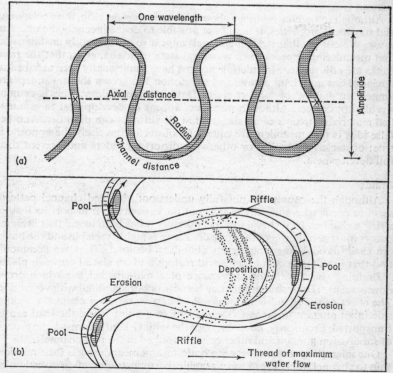

Fig. 5.7. (a) Meander geometry. (b) Erosion and deposition in a meander.

by their amplitude, wavelength and radius. Of these **wavelength** is very significant, especially in relation to the width of the river and discharge at bankfull stage. The wavelength is empirically related to the square root of the discharge, and it has also been established that wavelength is normally ten times the bedwidth.

A meandering river is asymmetric in cross-section at bends, the greatest depths occurring near the outer banks. Between bends, the bed of the river is shallower and more symmetric. The deeps at bends are known as **pools**, while the shallows between bends are **riffles**. Meandering appears to begin with the development of pools and riffles in straight channels, where the fastest water flow naturally swings from side to side. The fullest development of pool and riffle sequences occurs in meanders when the river is at or near bankfull discharge. This is why bedwidth on meandering channels is accordingly measured between banktops.

In a typical meander the surface water flows towards the outer bank, while the bottom water flows towards the inner bank. This corkscrew-like arrangement is known as **helical flow**. Helical flow accounts for the variation of channel profile in cross-section and for the promotion of local erosion and deposition in a meander (Fig. 5.7b). We find that **point bars** develop on the inside of bends and eroded **slip-off slopes** on the outer where water flow is fastest.

Although something is known about how meanders develop, their behaviour and statistical properties, it is still not possible to define precisely the ultimate cause. We can say that straight channels appear to be inherently unstable and that meandering appears to be a natural state of affairs, given that the river banks are adjustable. Meandering streams have traditionally been ascribed to fluvial plains and delta plains, and the notion has grown that meanders are limited to 'mature' or 'old' rivers of Davisian terminology. This is demonstrably not correct. Although meanders appear to develop best in granular bed materials, streams of all sizes and at all altitudes can meander. Another false idea is that meanders are caused by obstacles: precisely the opposite is true; obstacles, geological or otherwise, distort meanders and prevent their full development.

Braids

Although the causes are not fully understood, braided channel patterns seem to occur where the stream has not the capacity to transport its load in a single channel, be it straight or meandering. It has been found that braiding occurs most readily where the river discharge is highly variable and the banks are easily erodable, supplying an abundant bedload. Thus we commonly find that braiding occurs in semi-arid regions and on glacial outwash plains.

Braiding starts with the appearance of a mid-channel bar which grows downstream. The upstream end may become dry or stabilised with vegetation. The effect of the island is to localise and narrow the river channel on either side in an attempt to increase the velocity to a point where the load can be transported. Frequently, however, once the initial island is formed, the process of subdivision goes on and other bars are formed in the new channels.

One interesting contrast between braiding and meandering is their relationship to channel slope. Fig. 5.8 is a graphical demonstration of an investigation which shows that a line of demarcation can be drawn between meandering and braided channels which indicates, for a given discharge, that meanders occur on lower slopes than braiding. Or to put this another way, meanders will occur at a smaller discharge than braiding on a given slope. This may mean that some rivers can change from a meandering to a braided pattern with an increase in discharge.

We may conclude that the map plan of a river is all part of the steady state adjustment of a stream to environmental controls. Once again, like the cross-section and longitudinal profile, the channel pattern—whether straight, meandering or braiding—is determined by the response of the river to the discharge and load provided to it.

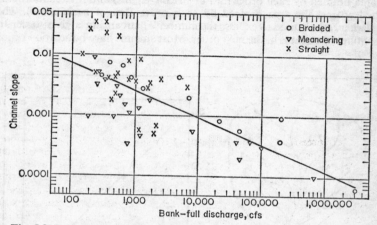

Fig. 5.8. Bankfull discharge plotted against channel gradient for a number of straight, braided and meandering streams.
(Courtesy US Geological Survey)

The River Basin

The geomorphologist is interested not only in the river itself, but also in its drainage basin, which includes all the area drained by the river and its tributaries. The drainage basin is bounded by a **watershed**, separating it from adjacent basins. There are several ways of classifying a river network on a descriptive or genetic basis, and examples of these will be considered here. However, in recent years considerable emphasis has been placed on the quantitative investigation of the geometric properties of rivers and their basins. This analysis is given the term **fluvial morphometry**.

Fluvial Morphometry

In order to facilitate comparisons between basins, a hierarchy of **stream orders** is used. This was originally proposed by Horton, but the amended systems of Shreve or Strahler (Fig. 5.9) are the ones generally in use. Streams without tributaries at the head of river systems are designated first-order streams. Two first-order streams join to form a second-order stream; two second-order streams join to make a third-order stream; and so on. If a lower-order unites with a higher-order stream, the order of the latter remains unchanged, i.e. it takes at least two streams of a given order to form a stream of the next higher order. The main stream is always the highest order in the basin, and the basin order is named from this. In Fig. 5.9a the example is a fourth-order basin.

The examination of a large number of systems has shown that if we count the number of streams in each order, then that number decreases with increasing order in a regular manner. This is known as the **law of stream numbers** and is demonstrated in Fig. 5.9b. The stream order is related to stream numbers on a semi-logarithmic scale, and a straight-line plot emerges. Similar straight-line plots can be obtained if stream order is plotted against the area drained by each order (**law of stream lengths**) and if stream order is plotted against the area drained by each order (**law of basin areas**). The **bifurcation ratio** is used to express the number of streams in a given order to the number of streams in the next order. Most streams have bifurcation ratios between 3 and 5.

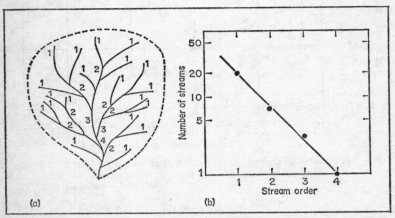

Fig. 5.9. (a) Stream orders. (b) Stream order plotted against stream number for the basin shown.

These various geometric properties show that in a given river basin, downstream changes are regular and determinable, so that laws of growth and development can be formulated, relating river length, number of tributaries, basin area, river gradient and relief to the order of the river. These laws are sometimes referred to collectively as the **laws of drainage composition**. Underlying them is a fundamental growth principle which in biological sciences is known as the law of **allometric growth**. This law, applied to an animal, states that the relative growth of an organ is a constant fraction of the relative growth of the whole individual. The same applies to the component parts of a river system in relation to the total growth of the system.

In addition to these laws, other useful indices can be derived from the analysis of drainage basins. **Drainage density** is the total channel length divided by the total area of the basin. In the United States, wide ranges of density values occur: the lowest, 2 to 2·5 km per square kilometre, are found on resistant rocks in the Appalachians; values of 30 to 60 are typical of the drier parts of the Rockies; and values in excess of 200 are reported from badlands. **Stream frequency** is the number of channels per unit area; like drainage density it is a measure of the texture of the drainage net and hence of the degree of dissection. The **constant of channel maintenance** defines the area required to sustain one unit length of channel. As such, one might be able to use it to predict whether there was any likelihood of the drainage system extending headwards.

Genetic Classification of Drainage Patterns

A genetic classification of streams was developed by W. M. Davis, following earlier ideas of J. W. Powell and J. B. Jukes. A **consequent** river is one whose course is determined by the original course of the land. Such streams can often be seen on a hillside, all flowing straight down the slope, forming a parallel drainage pattern. A **subsequent** stream is one whose direction of flow is controlled by rock structure; for instance, a river which follows a zone of weakness, such as a fault, joint or a line of weak rocks. A river is said to be **insequent** if there is no apparent reason why it follows the path it takes. Such rivers demonstrate a lack of structural control and tend to develop homogeneous rock. The patterns of these rivers tend to be dendritic, with tributaries branching in all directions. **Obsequent** streams flow in a direction opposite to that of the original consequent slope. **Resequent** rivers are those

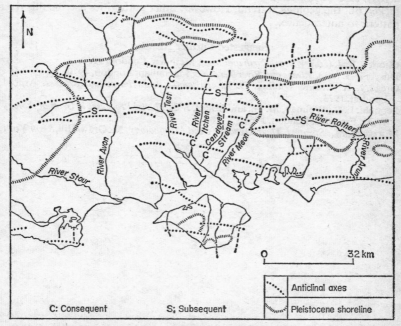

Fig. 5.10. The superimposition hypothesis for the origin of the drainage of central southern England.

which flow in the same direction as the consequent streams but at a lower level.

These terms have been widely applied to the streams developed on folded strata such as exist in southern England (see Fig. 5.10). By virtue of their definitions these terms carry considerable implications as to the history of erosion in any region. Their use is only justifiable if the erosional history of a region is known with some degree of certainty.

Other genetic terms applied to rivers relate to those streams which anomalously cut across geological structures such as folded ridges or crystalline mountain ranges. These rivers are said to be **discordant to structure** and are either antecedent or superimposed. If the course of the river predates the elevation of the mountain or ridge, and yet managed to keep its path, then it is **antecedent**. However, if the folded rocks were covered by horizontal layers upon which the river took its original course, and then the river eroded its channel down into the underlying ridges, then the river is **superimposed**. Again, the use of these terms implies that something is known about the history of denudation of an area. It is thought that many of the major English rivers such as the Trent and the Thames, which cut across varied geology, are superimposed, or **epigenetic**, to use an alternative term. Fig. 5.10 depicts the superimposed drainage of central southern England. The main north–south rivers (C) are thought by some authors to have been initiated on a covering layer of Pleistocene marine gravels, but now flow discordantly across east–west trending anticlines. Subsequent streams (S) have since picked out the geological trend. However, other authorities ascribe this river pattern to antecedence.

Suggested Further Reading

Chorley, R. J. (ed.), *Water, Earth and Man*, Methuen, London, 1969.

Dury, G. H. (ed.), *Rivers and River Terraces*, Macmillan, London, 1970.

Gregory, K. J., and Walling, D. E., *Drainage Basin Form and Process*, Arnold, London, 1973.

Leopold, L. B., Wolman, M. G., and Miller, J. P., *Fluvial Processes in Geomorphology* W. H. Freeman, London, 1964.

Morisawa, M., *Streams, Their Dynamics and Morphology*, McGraw-Hill, New York, 1968.

CHAPTER SIX

GLACIAL AND PERIGLACIAL LANDFORMS

Only about 10 per cent of the world's land surface is occupied by glaciers today; the ice-sheets of Greenland and Antarctica make up the great bulk of this amount. In the northern parts of Asia and North America large areas exist where freeze–thaw action is one of the dominant processes. This is the **periglacial** region: it approximately corresponds to the subpolar climatic zone (zone 6) described in Chapter Fifteen. During the cold phases of the Pleistocene, both the glacial and the periglacial regions expanded greatly, and left a strong legacy on the landforms of many temperate regions of the world, including the British Isles, New Zealand, southern Australia and much of North America. In this chapter we shall be mainly concerned with a systematic view of glacial and periglacial action. A consideration of the chronology of Pleistocene glaciation is reserved for Chapter Nine.

Ice and Glaciers

The study of the behaviour of glacial ice is known as **glaciology**. Glaciers by their very nature present problems for field investigation, especially within the glacier and at the glacier sole, where many important landform processes take place. We have sufficient knowledge, however, to say how glaciers work in outline.

All glaciers are derived from snow. When this first falls it has a density of 0·1, but it undergoes settling and compaction, gradually changing its density to become névé (firn) at 0·6, and it eventually transforms into polycrystalline ice at a density of 0·9. At this stage the glacier is made up of millions of ice crystals, each with well-defined facets.

Glaciers will only exist and grow if the annual **accumulation** of snow exceeds the annual **ablation**, or melting. Ablation occurs by radiation and by conduction when it rains in summer, the bulk of the melted snow being carried off as glacial meltwater. The relation between the amounts of accumulation and ablation is known as the glacier **mass budget**. The budget can either be positive (accumulation greater than ablation), or negative, or in balance. The mass budget determines whether the glacier will advance or retreat: there is usually a lag of several years between a change in the annual budget and the response of the glacier snout, and this enables a study of the net annual budget to be used to predict glacier behaviour. Fig. 6.1 shows a monthly plot of ablation and accumulation for one year.

In winter, when snow falls all over the glacier, the snowline lies well below the glacier snout. In summer, the snow on the lower part of the glacier melts, but remains in the upper reaches. The lowest point at which snow manages to remain each year marks the **summer firn line**, which delimits the two spatial parts of a glacier, an accumulation zone and an ablation zone (Fig. 6.2). The loss of material in the ablation half of the glacier is partly replenished by the downward movement of ice in the glacier. Glaciers can be regarded as an

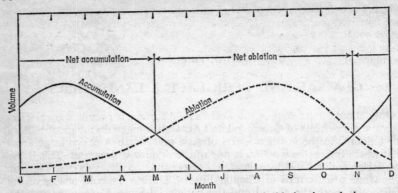

Fig. 6.1. The yearly course of accumulation and ablation in a glacier.

example of a cascading open system, moving energy and material as part of the larger hydrological cycle.

Glacier Movement

Glaciers move downhill under gravity. They can also move uphill at their bases, provided the surface slope of the glacier is generally downhill. The glacier acts partly as a solid, moving by sliding at the base and sides and by internal shearing, which expresses itself at the surface in the form of crevasses. It also acts partly as a plastic medium, deforming under stress by reshaping and realignment of its crystals.

Bedrock exerts frictional drag on the moving glacier and, as in the case of a river, the maximum rate of flow is near the middle of the ice stream (see

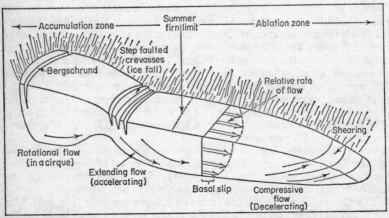

Fig. 6.2. Section through a glacier, showing types and rates of flow

Fig. 6.2). Most glaciers flow rather like a concertina: in some parts, **compressing flow** may be recognised, in which the shear planes thrust upwards. In other sections, **extending flow** takes place. Irregularities in the bedrock floor may initiate these patterns, but it seems likely that, in turn, the flow patterns accentuate the bedrock morphology (positive feedback). Compressing flow may favour erosion by moving material upwards away from the rock/ice interface.

Types of Glacier

Glaciers can be classified according to their morphology, their activity or their temperature. **Cirque glaciers** occupy amphitheatre-like depressions at high elevations. These glaciers may extend into **valley glaciers**. The coalescence of several valley glaciers on a plain creates a **piedmont glacier**. The further build-up of these topographically-guided glaciers creates **ice-caps** and ice-sheets which largely submerge the pre-existing landscape.

The degree of activity of a glacier can be expressed by its regimen. **Glacier regimen** is a function of three factors: the total mass budget, the state of the net annual budget and the rate of movement, itself controlled by the budget and by gradient. An **active** glacier can have a variable regimen, but the essential point is that forward internal motion is taking place; this applies irrespective of whether the position of the glacier snout is advancing or retreating. A **passive** glacier has a low regimen and little forward motion. A **stagnant** glacier has a chronic negative budget and is wasting away on the spot. Glaciers become stagnant because of climatic change, or because they have been separated by downwasting from their accumulation areas. The state of activity has an important bearing on the origin of glacial landforms: drumlins and arcuate end-moraines are formed by active glaciers; kames, eskers and other fluvioglacial features can only have been created in a stagnant environment, otherwise they would become flattened by an active glacier.

Glaciers are not all equally cold. A broad distinction can be made between cold **polar glaciers** which have very low temperatures throughout and are normally frozen to their bedrock, and so-called warm or **temperate glaciers**, which are much closer to 0°C and are not frozen to their beds. Fig. 6.3 shows typical temperature profiles for the two types. Note that in the temperate type, despite the gradual fall of temperature with depth, the increase of pressure means that the glacier can be near pressure melting point throughout.

Polar glaciers, frozen to the valley floor, will move largely by internal shearing and deformation, and will be unable to abrade their beds. Temperate glaciers on the other hand will have a high component of basal sliding in their motion, with much greater possibilities for erosion and basal deposition. Recent evidence suggests that these contrasting thermal regimes may exist in different parts of the same glacier. Antarctica, for example, appears to have a polar regime at its periphery, but towards its centre the tremendous weight of the ice causes pressure melting and allows unfrozen bodies of water to exist at the base of the ice.

Glacial Erosion

Glacial erosion processes, taking place at the sole of the glacier, are nearly all mechanical effects. Chemical erosion is of little, if any, significance. Two principal processes are regarded as important: abrasion and plucking.

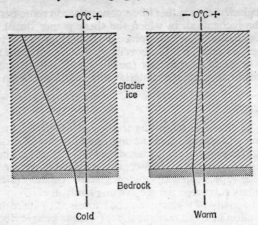

Fig. 6.3. Temperature profiles through cold (polar)
and warm (temperate) glaciers.

Abrasion, the grinding and crushing of rocks, is not accomplished by the ice itself as it is too soft, but by the rock debris frozen into the lower layers of the ice. The debris load acts rather like a coarse sandpaper, and also abrades itself as it is moved by the glacier. **Plucking** or quarrying occurs in response to the drag exerted by the moving ice on the bedrock. It is not a simple tug-of-war, because the tensile strength of ice is not very great; plucking is most effective where the rock has already been weakened by being well jointed, or by being subject to pressure-release and freeze–thaw action beneath the glacier.

Plucking is quantitatively more important than abrasion and produces a fretted land surface and probably contributes much to the morainic debris carried by glaciers. Abrasion has a smoothing effect and creates rock flour, which gives meltwater streams their characteristic milky appearance.

Small-scale Erosion Forms

The smallest-scale forms of erosion are scratches, grooves and various other markings created by abrasion. **Striations** are the most common of these, and are fine lines inscribed on bedrock by an angular boulder carried in the basal ice. Crescentic fractures, cracks and gouges are sometimes collectively known as **friction cracks** and are concave downstream. They are thought to be produced as the bedrock cracks under the oblique pressure of a moving boulder. All these small abrasion features are useful indicators of the direction of ice movement, although striations can of course indicate two possible directions, and supporting information is usually required.

On a larger scale, **roches moutonées** (Fig. 6.4a) also give clear evidence of passing ice. These are small asymmetrical rock knobs which have abraded stoss (upstream) sides and plucked lee sides. This change of process within the one feature may be related to small irregularities in the topography of the bedrock floor, which are picked out by the ice, or to the jointing pattern in the rocks (Fig. 6.4b). Irregularities in the bedrock floor are subject to greater ice pressure on their upstream sides than on their lee sides. On the stoss side this causes pressure melting, and the ice abrades by sliding. On the lee side, refreezing of the glacier to the bedrock allows plucking to take place. Similarly in a situation of variable jointing, where the rock is well jointed, plucking is favoured; on the less well jointed stoss side, the rock surface is moulded by abrasion. In this way, glacial erosion accentuates pre-existing landforms.

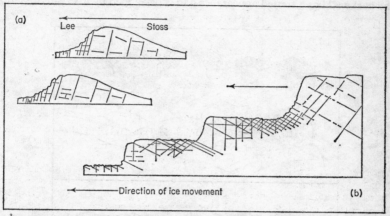

Fig. 6.4. Roches moutonées (*left*) and rock steps related to jointing.

Large-scale Erosion Forms

Large-scale erosion forms include U-shape valleys, hanging valleys, cirques and arêtes (Fig. 6.5). They are characteristic mainly of high relief regions, where considerable potential energy exists for glacial erosion. The well-known **U shape** of a glaciated valley is formed because erosion occurs throughout the base of the glacier, becoming progressively less at the sides, rather than being concentrated at one point, as in a river valley. Well-developed glaciated valleys are known as **glacial troughs**: at their upper end they may have a steep head wall, known as a trough end. Above the headwall, several tributary glaciers may have fed into the main trough. Glaciated troughs typically possess truncated spurs, rock steps and basins. **Rock steps** are similar to, although much larger than, roches moutonées and have abraded lips and plucked lee sides: again, steps may be related to jointing (see Fig. 6.4) or to resistant lithologies. Alternatively, steps may develop where two or more glaciers meet: the extra volume of ice will result in an increase in erosive power. Rock basins are sections of the valley overdeepened by the glacier by rotational movement in the ice, and frequently develop where the ice surface gradient is steep. Many glaciated rock basins now contain lakes or are inlets of the sea (fjords), as in Norway, New Zealand and western Scotland.

Where a main glacier was formerly joined by a smaller tributary, a **hanging valley** may be created. The relative size of the two valleys is directly related to the power of erosion of the glaciers occupying them. In the reverse situation, where a powerful trunk glacier overspills the valley confines, thus breaching the old watershed, a **glacial breach** is formed. On deglaciation, this will be left as a broad col connecting one highland valley to the next.

Cirques (corries, cwms), when fully developed, have an amphitheatre-like form with steep backwalls and small rock basins often occupied by lakes after the ice has disappeared. The basic form arises from the rotational movement in the ice induced by the steep surface gradient of the glacier (Fig. 6.6). Abrasion appears to be the principal process which excavates the basin, but the backwall has a fretted appearance and is developed by freeze–thaw action

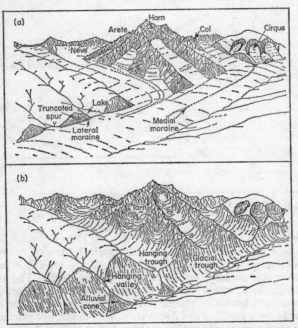

Fig. 6.5. Some features of mountain glaciation, (a) during and (b) after glaciation.

(After A. N. Strahler, courtesy J. Wiley & Sons)

in the **bergschrund**, a deep crevasse at the back of the glacier. The headward coalescence of adjacent cirques on a mountain mass may cause their sides to intersect and form **arêtes** and **pyramidal peaks**.

All cirque glaciers develop from snow banks. Once the snow has become firn, the process of **nivation** (erosion by snow) starts to deepen the hollow through diurnal freezing and thawing. Some cirques do not proceed beyond the firn stage, and are known as **nivation cirques**. These do not possess rock basins or moraines, but may have frontal mounds of **protalus** material, built by debris sliding down the surface of the snow patch. For those that develop into true glacial cirques, we can see something of their snow-bank origins in the fact that in Britain many of them are located on the north-east sides of mountains. This would have been the most favourable location for snow

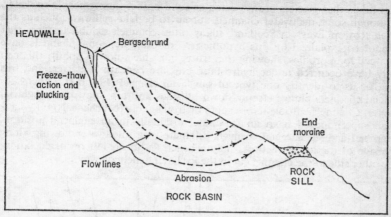

Fig. 6.6. A cirque glacier in cross-section.

accumulation, in the lee of the prevailing wind, and partly shaded from maximum insolation.

Erosion in a Lowland Area

Erosion by large ice-sheets of lowland areas produces a scoured, generally debris-free landscape, except near the maximum limit of glaciation where end-moraine deposition occurs. On old, hard rocks glacier ice picks out rock knobs and depressions: in north-west Scotland, glaciation of the Lewisian gneiss has produced a landscape known locally as **knock-and-lochan** topography.

Many observers have pointed out that in some areas known to have been long occupied by Pleistocene ice-sheets, relatively little erosion seems to have been accomplished. This applies notably to large areas of the Canadian Shield. Ice is not necessarily always a powerful erosive force and in some cases it can have a relatively protective role. This variation in erosive effect is known as **glacial selectivity**. The capability of ice to erode depends largely on the thermal characteristics of the ice and the gradient over which it is flowing. A sluggish polar-type glacier in a flat lowland area may therefore achieve little erosion in quantitative terms, whereas a temperate-type glacier flowing down the relatively steep gradient of a mountain valley will be much more effective.

Meltwater Erosion

Meltwater is an important feature of all glaciers, and moves in tunnels subglacially (under) and englacially (within), before emerging to flow along the side of the ice or downvalley away from the glacier snout. Meltwater highly charged with rock and ice debris and moving under great hydrostatic pressure may be capable of moulding solid rock into smooth, round forms. However, the main erosional features associated with meltwater are **meltwater channels**, steep-sided and flat-floored valleys not unlike railway cuttings.

It used to be thought that all meltwater channels were eroded by the overspill of water from glacially dammed lakes. The erroneous assumption was made that glaciers always acted as solid bodies, impermeable to water. Although some meltwater channels appear to be **lake spillways**, such as the Glen Roy spillways in Scotland, many other channels exhibit features that cannot be explained by this hypothesis. For example, some channels have humped long profiles, showing that water must have flowed uphill; this can only have occurred under hydrostatic pressure beneath a glacier, and this enables us to identify one type of **subglacial channel** (Fig. 6.7). Other subglacial channels plunge steeply down hillsides and are known as **subglacial chutes**. Channels cutting across spurs or trough cols (**col channels**) may have been superimposed from an ice-tunnel originally in an englacial position. **Marginal and submarginal channels** originate from meltwater flowing along the side of the ice. Sometimes only a bench feature is left on deglaciation, the other side of the channel cut in the ice having melted away.

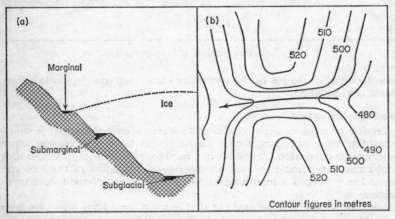

Fig. 6.7. Meltwater channels: (a) subglacial, submarginal and marginal positions; (b) contour map of a spur channel.

Many hillsides and valley floors in northern England, Wales and Scotland possess glacial meltwater channels. Those channels cut into hillsides are usually dry and play little part in the present drainage pattern. Lowland channels are more likely still to be used by modern rivers, now misfit in the larger meltwater form, as exemplified by stretches of the River Irwell in Lancashire.

Glacial Deposition

The great bulk of the material eroded by a glacier remains in the lower layers of ice and is soon redeposited. However, some material works its way upward through the ice on shear planes to emerge at the glacier surface and then to be carried supraglacially. Sometimes boulders are transported great distances; such **erratics**, moved far from their original bedrock, are useful indicators of the direction of ice movement. In addition to direct ice transport, meltwaters may pick up debris and redeposit it. Hence a twofold division of glacial deposits is usually recognised: material deposited directly by ice is known as **glacial till**; that deposited by meltwater streams is referred to as fluvioglacial material. The general term **glacial drift** is given to all debris deposited in a glacial environment.

Till Deposition

Till is a very heterogeneous deposit composed of a seemingly chaotic assemblage of stones of various sizes set in a finer mass (matrix) of clay, silt or sand. Till is sometimes known as **boulder clay**, but many tills have neither boulders nor clay. The individual stones in till are usually subangular—that is, they are not rounded, but neither do they possess sharp edges. The lithology of the till will reflect the ground over which the glacier has passed: in mountainous areas, tills tend to be rather rubbly; tills in East Anglia are chalky, having passed over the chalk outcrop; around the shores of the Irish Sea, many tills are clayey and contain shells, dredged from the sea floor.

Lodgement till is till that has been plastered on to the ground by the sole of the glacier, perhaps against an obstruction. The lower debris-rich layers of ice become separated from faster-flowing and clearer ice above. The forward motion of the glacier causes the till stones to be aligned so that their long axes are parallel to the direction of flow. A careful examination of the alignment of stones in old till deposits can be a guide to the former direction of flow of the ice-sheet. When a stagnant glacier downmelts on the spot, englacial and supraglacial material will accumulate in heaps as **ablation till**. In this case we would not expect the stones to be aligned in any particular direction. Material brought to the ice surface on shear planes sometimes flows rather like porridge, filling up any hollows on the glacier surface and forming a **flow till**. Hence with these several different possibilities for till deposition, a single glacier can build up quite complicated drift sequences.

Moraines are depositional landforms, usually composed of both till and fluvioglacial material. Lodgement till normally makes up most of the **ground-moraine** of the glacier, creating a fairly featureless area or plain, unless drumlins (see below) are also present. At the margins of the glacier, more definable ridges or mounds are created as part of the **end-moraine complex** (Fig. 6.8). The **lateral moraine** builds up at the side of the glacier, the **terminal moraine** at the glacier snout. End-moraines take a wide variety of form and usually contain fluvioglacial material as well as till. The largest end-moraines may reflect a long standstill of the glacier margins in one position, and may contain multiple ridges piled up by a slight readvance of the ice over previously deposited drift. **Recessional moraines** mark standstill stages in the retreat of a glacier.

Drumlins are streamlined egg-shaped features, largely composed of lodgement till, although some may have a rock core. Individual drumlins can be up to a kilometre in length and they usually have their long axes aligned parallel to the direction of flow; their steepest (stoss) end faces up-glacier. The expression **basket-of-eggs topography** arises because drumlins usually occur in clusters. Some theories suggest that they are erosional features, formed by the ice-moulding of drift deposited in a previous glaciation; other theories suggest that the moulding is imparted as the till is being laid down.

Fluvioglacial Deposition

The main characteristic of fluvioglacial drift is that it has been sorted and stratified into layers by the meltwater. However, a glacial environment is such a variable one that this stratification often changes rapidly over short distances. Two types of material can be distinguished: ice-contact drift, deposited against ice; and proglacial drift, carried beyond the ice-margins.

Ice-contact drift accounts for several distinct landforms, some of which are shown in Fig. 6.8 above. **Eskers** are long narrow ridges of well sorted material deposited in sub- or englacial tunnels. **Kames** are individual mounds of material of various origins: some kames may represent former crevasse fillings, others may be the remains of small deltas. Kame deposits frequently exhibit faulting, formed as the material slipped when the ice melted away. **Kame terraces**

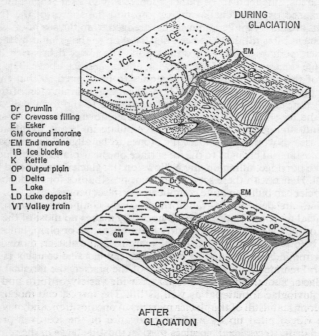

Dr Drumlin
CF Crevasse filling
E Esker
GM Ground moraine
EM End moraine
IB Ice blocks
K Kettle
OP Output plain
D Delta
L Lake
LD Lake deposit
VT Valley train

Fig. 6.8. Depositional features, during and after glaciation.

accumulate in the area between the side of the glacier and the valley walls (Fig. 6.9). They have a roughly level surface controlled by the height of the local englacial water table, and usually include rafts of till and scree material as well as fluvioglacial deposits. **Kettle holes** are small enclosed depressions formed by the melting of partially buried blocks of ice, usually within fluvioglacial material. Areas of mounds and depressions commonly associated with a stagnant glacier are referred to as **kame-and-kettle topography**.

The main landform created by proglacial deposition is the **outwash plain** or **sandur**. When the material is confined to a valley, the term **valley train** is used. These features are built up by the constantly shifting meltwater streams, dumping the coarsest material at the proximal end near the glacier margin, and carrying fine material to the distal end of the plain or train. The retreat of a glacier may result in several outwash features being formed, each related to a recessional moraine.

Where proglacial debris is deposited into a lake, a delta will form. The fine bottom-set material deposited in the middle of the lake may be varved. Each **varve** consists of a pair of laminations, a coarser one representing summer deposition, and a finer one being the result of the slow winter precipitation of the finest material in the lake. The counting of varves has been used as a method of dating events.

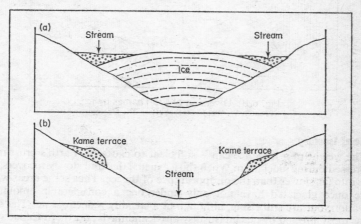

Fig. 6.9. Formation of a kame terrace.

Other Effects of Glaciation

Drainage Diversion by Ice

Glaciation has important effects on the landscape beyond the direct modifications created by ice erosion and deposition. One example is that glaciation frequently disrupts pre-existing drainage lines, initiating a new pattern persisting after the ice has disappeared. Two well-documented instances occur in England. In the Midlands, what is now the Coventry/Warwick area was formerly drained by the headwaters of the River Soar, flowing into the Trent south of Nottingham. During glaciation, the valleys of the Soar and Trent were occupied by ice advancing from the north. At the same time, ice advanced from the Welsh mountains into the Vale of Evesham and combined with northern ice to pond up a large proglacial lake covering much of the Midlands. During deglaciation, Welsh ice retreated first, allowing the lake waters to escape westwards, initiating a new river, the Warwickshire Avon. The drainage pattern of the Midlands was hence reversed from one flowing north-east towards the Humber, to one now flowing into the Bristol Channel.

The second instance is that of the Thames. Successive courses of the Thames are shown in Fig. 6.10. On each occasion, the diversion southward to a new course seems to have been promoted by glaciers advancing into the Vale of St. Albans and later into the Finchley depression.

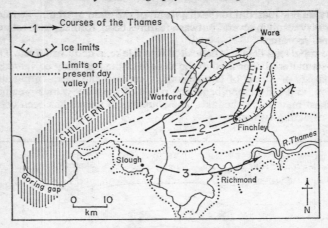

Fig. 6.10. Diversion of the Thames by ice.
(From S. W. Wooldridge, courtesy The Geologists' Association)

Glacial Isostasy

The weight of a large ice-cap is sufficient to cause the Earth's crust to be depressed during glaciation. When the ice melts, the land slowly recovers, but at a rate far slower than the disappearance of the ice. Thus some areas which were once glaciated have not only undergone a considerable amount of elevation, but are still rising, although at a steadily reducing rate. The three areas most affected by glacial isostasy are Scandinavia, northern Canada and northern Britain. A great deal can be gleaned about the recovery of these areas from the study of raised beaches (Chapter Eight) and other shoreline phenomena which were formed by seas abutting the landmass as it recovered.

Lines joining points of equal recovery are known as **isobases**. The greatest amount of recovery in each region takes place near the centre of glaciation, where the ice was thickest. The isobases form a concentric pattern around this area and decrease to zero near the limit of glaciation. In Britain, the zero isobase runs approximately from the Humber to North Wales. The greatest amount of recovery anywhere in the world is recorded by Baltic raised beaches at a height of $+275$ m; this can only represent part of the total amount, since a great deal of recovery must already have taken place before the first raised beach was formed.

Periglacial Phenomena

Periglacial regions are very active geomorphological areas. The mechanical splitting of rocks by ice (gelifraction), frost heaving of the ground (geliturbation), solifluction and nivation are all important processes. In addition, each spring, large quantities of water from melting snow and ice rapidly erode the debris scattered and moved down the slopes. Wind action is also a significant force.

Basic to several processes in periglacial areas is the formation of **ground ice** within the soil and upper bedrock layers. Ground ice is significant because, as it forms, it tends to segregate into needles, lenses or veins, thus causing much more local disruption than would be achieved by the normal ten per cent increase in volume. The freezing of moisture in soil is also capable of disrupting the colloidal attraction between soil particles, on melting rendering the soil layers incohesive and mobile. Not all sediments are equally liable to disturbance by frost action: **frost susceptible materials** are mostly in the silt-size range.

In some parts of the periglacial zone, geomorphic activity is promoted by a large number of freeze–thaw cycles and high precipitation amounts (Icelandic type of region). Other areas (Siberian type) are characterised by light winter precipitation, few freeze–thaw cycles and very low temperatures, resulting in the presence of **permafrost**, the name given to frozen ground. Permafrost consists of a permanently frozen layer, reaching to considerable depths, and an **active layer** nearest the ground surface which unfreezes in summer. This arrangement produces some unique effects, as explained below.

Patterned Ground and Involutions

Repeated freeze–thaw cycles and the breaking and heaving of the ground have the effect of differentially sorting the finer frost-susceptible materials away from the coarser particles in the soil. In plan view, this produces patterned effects on the ground surface. On flat ground, individual **stone circles**, or nets of **sorted polygons** may form, each with a coarse border and fine centre (Fig. 6.11a). On steeper ground, the polygons elongate into garlands and **stone stripes** form on gradients of 7° or more. Many of these features can be observed forming today near the summits of the higher mountains in Britain.

Where the soil is more uniform, and the sorting of material is less obvious, the ground may be heaved into small mounds or **earth hummocks (thufurs)**. This is sometimes referred to as unsorted patterned ground.

Involutions (Fig. 6.11b) are contorted vertical structures, usually only seen in fossil form in pits and quarries. Their origin in some cases is connected with patterned ground in that differential heaving occurs in the form of

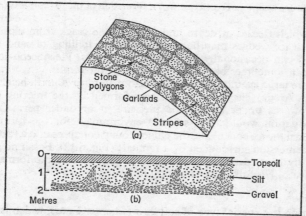

Fig. 6.11. Periglacial phenomena: (a) patterned ground;
(b) involutions.

regular festoons. Another possibility for the formation of involutions is that they are the result of the pressure put on the active layer in permafrost when it begins to refreeze from the surface downwards. Periglacial involutions are sometimes difficult to identify positively, as load structures and solution pipes often have a similar appearance. Patterned ground and involutions can have important effects on tundra vegetation, but they are not significant land-form features.

Ice Structures

Ice wedges are vertical structures, two or three metres in depth, found in active form only in permafrost regions. The intense cold causes the ground to contract and crack into polygonal networks (**unsorted polygons**). The initial crack becomes wider each year because in the summer it is filled with

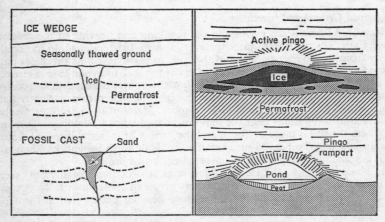

Fig. 6.12. Ice wedge and fossil cast (*left*); formation of pingo ramparts (*right*).

moisture which freezes at depth and prevents the crack from closing (Fig. 6.12). Fossil ice wedges may be preserved by an infilling of sand or other debris, and can occasionally be seen in gravel pits. Pleistocene ice-wedge polygons can sometimes be spotted on aerial photographs.

Pingos are large ice mounds, anything from 10 m to 70 m in height, found in the high Arctic. They have an ice core formed by the freezing of water moving upwards under hydrostatic pressure in or under permafrost. An artesian situation would be a suitable environment for the formation of pingos. When the ice core melts, the pingo mound collapses into a feature with a central depression surrounded by a rampart (Fig. 6.12). Fossil pingos have only recently been identified in Britain, but groups of these forms exist at Walton Heath, Norfolk, and Llangurig in Wales.

Solifluction

Solifluction is a special type of soil flow (Chapter Four), which is particularly widespread in permafrost regions. This is because the soil layers are rendered incohesive by the effect of freezing and thawing, and because the permanently frozen ground prevents the downward percolation of water in summer, producing a highly saturated and mobile soil layer. In addition, there is no deep-rooted vegetation in tundra regions to bind the soil. Solifluction in permafrost areas can occur on slopes of 3° or less, and has the effect of removing waste from interfluves and filling up the valleys and hollows, giving a smooth appearance to the landscape. In the Pleistocene this type of solifluction appears to have been particularly widespread in the Chalk areas of southern and eastern England, choking the valleys with **coombe rock** and contributing to the rounded appearance of the outcrop. In other areas, solifluction debris is also known as **head**.

Solifluction occurs in periglacial regions without permafrost, under conditions of substantial precipitation and fairly steep slopes. Here, solifluction forms small arcuate stone-banked or turf-banked terraces and lobes, and these can be observed on some British mountains.

Slope Features

Some of the processes we have discussed earlier in this chapter, such as nivation and solifluction, clearly contribute considerably to the modification of slope form in periglacial environments. Several processes often interact to produce specific features.

At outcrops of hard rock, the combination of frost-shattering and removal of the debris by solifluction produces angular free faces and **tor-like forms** where the joint structures intersect at right angles. These tors can be similar in appearance to tors produced by deep chemical weathering in warmer environments of the globe. **Blockfields** may accumulate at the foot of particularly massive free faces. **Altiplanation terraces** are related features (Fig. 6.13), being semi-horizontal surfaces or benches cut into the hillside by a combination of frost-shattering, nivation and solifluction. Some terraces may have originated as a small hollow, perhaps occupied by snow-patches.

Where a hard competent rock overlies a band of clays, the freezing and thawing of the moisture-bearing clays may cause them to become mobile and

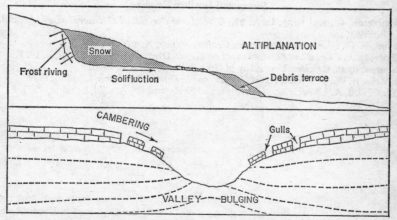

Fig. 6.13. Periglacial slope processes: altiplanation and cambering.

flow under the weight of the overlying rocks (Fig. 6.13). This in turn causes downslope **cambering** in the competent rock above, and the creation of **gulls** or clefts, as it breaks up into detached portions. In the valleys, the soft clays may be squeezed up to form **valley bulges**, which will be later partly eroded by stream activity. All these features are well developed in the central Weald of Kent and Sussex and also on the clays and ironstones of Northamptonshire.

Wind Action

The disturbed ground and lack of vegetation allows wind to become an effective landform agent in periglacial regions. This particularly seems to have been the case in the Pleistocene cold stages, since we now find extensive spreads of wind-blown **loess**, stretching in a broad zone from western Europe through Asia into China. In Britain, the more limited deposits of this fine material are known as **brickearth**. The loess appears to have originated by the deflation of glacial outwash plains, dry sea beds, and the tundra itself. Deposition normally creates fairly featureless landscapes, but occasionally dune forms are found. Periglacial wind action was also responsible for shaping stones by wind blasting into **ventifacts**, often found in Cheshire in association with outwash sands and gravels.

Fluvial Action

River flow in the Arctic periglacial zone tends to be characterised by irregular and sudden fluctuations. In winter, there may be a complete cessation or greatly diminished flow under ice. Rapid thaw in spring produces high discharges and flooding. Snowmelt and solifluction provides the channel with large amounts of debris, leading to the development of braided sections. On minor streams, jamming of debris causes **surges**, which helps to explain the extraordinarily large size of material found on the beds of some courses. However, not all Arctic rivers are of this character. In low-lying areas underlain by continuous permafrost, lakes, marshes, sluggish or disappearing streams are common.

Suggested Further Reading

Embleton, C., and King, C. A. M., *Glacial and Periglacial Geomorphology*, Arnold, London, 1968.

Flint, R. F., *Glacial and Quaternary Geology*, Wiley, New York, 1971.

Price, R. J., *Glacial and Fluvioglacial Landforms*, Oliver & Boyd, Edinburgh, 1971.

Sharpe, R. P., *Glaciers*, University of Oregon Press, Oregon, 1960.

Sparks, B. W., and West, R. G., *The Ice Age in Britain*, Methuen, London, 1972.

Washburn, A. L., *Periglacial Processes and Environments*, Arnold, London, 1973.

CHAPTER SEVEN

DESERT AND TROPICAL LANDFORMS

This chapter considers the landforms of the subtropical deserts of the world, their semi-arid margins, savanna lands and tropical rain-forests. There is no rigid distinction between arid and humid landforms in the tropics, although the relative intensity of different processes will obviously vary with the moisture change. The climatic transition is a very gradual one, and the intermediate savanna zone appears to experience landform processes more widely found in the arid and humid extremes. In addition, the picture is complicated by the possibility that Pleistocene climatic change has left relict arid landforms in the now more humid parts of the tropics and vice versa. Much of our knowledge of these landforms comes from observations in Africa and the deserts of North America and Australia.

Deserts

There is a popular misconception that arid landscapes are dominated by great seas of sand dunes, and therefore, it is argued, wind action must be the paramount geomorphological force, and water action of minimal importance. These notions need some correction; there is a great variation of relief and structure in deserts, large dune areas (**ergs**) occupying only ten per cent of the Sahara. Many other areas are mountainous, or characterised by almost featureless gravel plains, sometimes with a thin scrubby vegetation in the semi-arid margins. The conventional climatic definition of a desert is a region which receives less than 25 cm (10 in.) of rainfall (Chapter Fifteen), but very few areas receive no rain; most places receive some precipitation at infrequent intervals in the form of heavy downpours. The relative roles of wind and water constitute a useful underlying theme in a discussion of arid landforms.

Wind Action

Wind moves sand in suspension, by surface creep, and by **saltation**—that is, in a series of hopping movements. The sand moved by these processes is remarkably uniform in size, between 0·1 and 0·5 mm. Particles smaller than this are carried considerable distances, perhaps to be deposited well beyond the desert areas. Particles larger than this, coarse sand and gravel, remain as a **lag deposit**. The trajectories of sand grains undergoing saltation (Fig. 7.1) are very significant in building dunes and similar features. Turbulent wind flow near the ground creates a variety of short, steeply sloping ascents, followed by a much more horizontal path before the grain hits the ground. The impact of the grain may cause other particles to move by creep.

The erosive effects of wind in deserts is of two distinct kinds. **Deflation** occurs with the initial removal of the sand-size products of weathering. Where this is confined to a specific area, perhaps where chemical weathering has been strong around a salt pan, a **deflation hollow** may be created. The deflation of larger areas leaves behind a gravel or stone-strewn surface known in

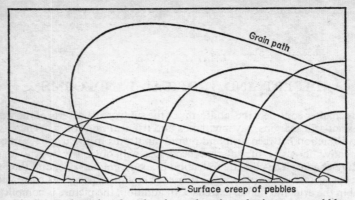

Fig. 7.1. Trajectories of sand grains undergoing saltation over a pebble surface.

the Sahara as a **hamada**. The other type of wind erosion is **abrasion**; this is confined to the saltation zone, which is normally within a metre or so of the ground. Wind abrasion is responsible for the undercutting and fluting of rocks, polishing hard rock surfaces, and shaping individual stones into faceted forms. Soft rock may sometimes be fashioned into **yardangs**, parallel U-shape troughs separated by sharp ridges. The extreme mushroom-shaped features occasionally observed in deserts are rarely the sole result of wind undercutting and involve chemical weathering also.

The depositional features created by winds in deserts include small-scale sand ripples and ridges, and large-scale dunes. The formation of **ripples** is closely connected with saltation. Most sand surfaces are initially uneven, but will be regularised in the following manner. Surfaces facing the wind (AB in Fig. 7.2) will receive more bombardment from saltation than lee slopes (BC). Material creeps up to B faster than it can be removed, and creeps away from A and C faster than the supply down the sheltered lee slope. The crest at B thus grows and the hollows deepen at A and C until an optimum regular ripple amplitude is developed in equilibrium with the prevailing wind speed. An equilibrium is reached because the crest at B will not grow indefinitely, as with increasing height it interferes with the wind flow and grains are blown off the crest into the hollow. Ripples will obviously not develop under very low or very high wind speeds.

Sand ridges are larger features than ripples and are composed of coarser material, too large to be lifted by the wind. The material in this case is shaped

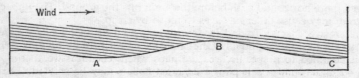

Fig. 7.2. The formation of sand ripples (see text).

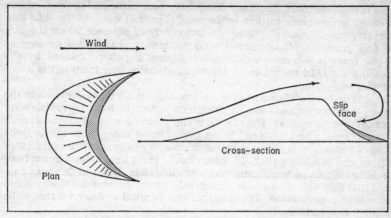

Fig. 7.3. Plan and cross-sectional view of a barchan dune.

into ridges by the impact of grains undergoing saltation, which can move coarse material up to six times their own size.

Although there are many shapes and sizes of dunes, two basic types can be recognised. **Barchan dunes** (Fig. 7.3) are formed from mounds of sand and become crescentic in plan with their horns pointing downwind. Material is moved by the wind up the windward slope of the mound and accumulates near the top of the lee slope, making that slope steeper. When this becomes too steep, it shears, the material slipping to the bottom of the slope. The process is repeated, and the dune slowly migrates. The horns are created because the process operates faster on the lower flanks of the mound; the rate of advance is inversely proportional to the height of the slip face. **Seif** (longitudinal) dunes are long and straight, often occurring in sub-parallel lines. Their

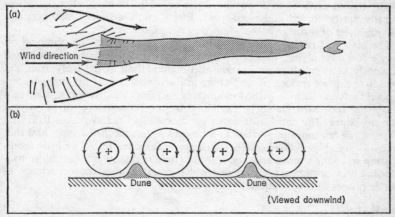

Fig. 7.4. Longitudinal (seif) dune: (a) formed in the lee of a hill; (b) formed by longitudinal roll vortices.

origin is problematical. Some seif dunes have been explained in terms of one dominant airflow: roll vortices in the wind (Fig. 7.4) whip up the sand into sharp ridges. Alternatively, there is a large body of evidence which suggests that seifs require multidirectional wind patterns, and that they develop as oblique features where a cross-wind elongates a pattern created by the prevailing wind. In special cases, longitudinal dunes may form in the lee of obstacles such as hills.

On a broad scale, if one plots the orientation of all types of dunes in the Sahara, their orientation shows a pattern consistent with the prevailing wind direction over large areas. The specific occurrence of groups of dunes depends on suitable dune-building sand being available. In some cases this may be within old basins of centripetal drainage; in other cases, near to outcrops of easily eroded sandstone rock. In Australia the great longitudinal dunes have a peculiar sinuous pattern, rather like a braided river. It has in fact been suggested that they are related to ancient dried-up river courses, for this is where sand will be concentrated. The principal role of wind in deserts seems to be to rework material locally.

The Role of Water

One of the most important effects of moisture in deserts is its role in promoting chemical weathering. This aspect has been discussed in Chapter Three. The effects are sometimes subtle but it is clear that a great deal of the weathering in deserts cannot be accomplished without the presence of some moisture in the atmosphere.

Water **run-off** has more dramatic effects. Because long periods of drought exist between the intermittent bouts of rainfall, weathering and wind action create large amounts of loose surface debris. Flash floods, caused by heavy convectional storms, rapidly pick up an enormous load with which they can accomplish a great deal of erosion. Initially under these conditions, run-off will be very great, with little or no vegetation to check it, but streams will eventually suffer heavy percolation losses and choke on their own debris. Where the water flows in a specific course or **wadi**, the channel is characteristically flat-bottomed and steep-sided. Elsewhere, **sheetfloods** may occur, a moving sheet of water over the whole desert surface.

The action of running water in arid regions may perhaps be observed on only a few days a year, and hence the tendency to underplay its role. It will obviously accomplish more in semi-arid regions than in the truly arid. Yet even in the driest regions of the Sahara and elsewhere, there are many water-eroded features, such as gullied mountains and deep wadi systems, that seem altogether too extensive to be explained by the present level of activity of running water. The conclusion many geomorphologists have come to is that water was even more important as a landscape former in the past, and that many topographic features we see now are undergoing relatively little modification in today's more arid conditions. There is a large amount of archaeological and documentary evidence which confirms the past occurrence of wetter phases in the Sahara and elsewhere.

Landforms

Dune areas and wadi systems are significant landforms in deserts, but over large areas, more typical is a landscape of vast flat or low-angle surfaces, broken sharply by isolated hills, scarps or mountain fronts. Some of the principal elements contributing to this landscape are shown in the landform assemblage in Fig. 7.5.

The upper part of the flat plane is composed of a rock-cut surface or **pediment**, which is succeeded downslope by an aggradational feature known variously as a **bahada** or **peripediment**. This in turn leads to either a river bed in semi-arid regions, or to a **playa**, an area of dried-up lake sediments dominated by salt derived by evaporation. The whole slope from the mountain front to the playa is called a **piedmont**. The rise of the scarp above the pediment is denoted by a sharp knick; this may locally be obscured by an alluvial fan, but elsewhere the knick is a very characteristic feature. The angle of the scarp is often the same in similar lithologies, whatever the size of the hill, strongly suggesting that parallel retreat is operating here.

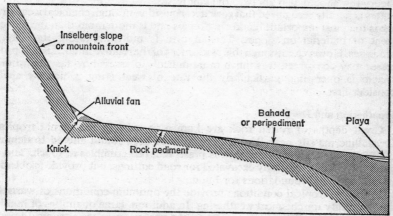

Fig. 7.5. Landform assemblage in a desert or semi-desert region.

One of the most problematical features of this assemblage is the origin of the rock-cut pediment and the sharp break of slope between it and the hill. Careful measurements of the pediment show that it has a slight concave-up profile, indicating that some kind of fluvial action is involved in its formation. While the precise processes of formation are not fully understood, it has been suggested that the pediment is cut by rill action and sheetwash, and that the scarp recedes by parallel retreat under the processes of weathering and mass movement. The sharp knick therefore records a specific change of process. In maintaining this break of slope there must of course be a delicate balance between the material falling down the scarp and its removal by the fluvial action at its base. In some cases localised weathering may also sharpen the knick.

This mechanism of pediment formation is called **pediplanation**, and this process has been strongly supported by the South African geologist, Lester King, as being responsible for the vast flat surfaces of Africa, including the veldt of southern Africa and the savanna lands farther north. Arising out of these surfaces are residual hills or **inselbergs**. They take various forms, depending on lithology, and may survive either because they are particularly resistant, or because of their position as the last unconsumed remnants of pediplanation. Lester King has suggested that pediplanation accounts for all inselberg forms in Africa, even in the rain-forest areas, but there are other possible explanations for inselbergs, as we shall see in the latter part of the chapter.

The Humid Tropics

The humid tropics embrace both tropical rain-forest and the seasonal rain-forest or savanna areas. The climatic and vegetational characteristics of these regions will be discussed in later chapters. Tropical rain-forest in particular is often regarded as the ideal natural system, in which there has been neither marked climatic change nor gross interference by man. Climate, vegetation and soils exist in a long-maintained dynamic equilibrium, and in theory this should also apply to the landforms.

It is generally recognised that **rock decomposition** through chemical weathering is the most important landform process and is more rapid than the movement of material on slopes. This in turn is more important than river processes. However, although this assessment of the relative importance of processes may be correct, it is much more difficult to assess how fast the whole system is operating, particularly the rate of weathering, which we shall consider first.

Weathering and Duricrusts

Great depths of rotted rock are known to exist in the humid tropics. Crystalline massifs near Rio de Janeiro have been found altered to depths of 100 m; the texture of the rock is preserved, but crumbles at touch. These great thicknesses are easily excavated for road cuttings, but provide problems in finding sound foundations for buildings.

In theory, tropical conditions provide the optimum conditions of warmth and humidity for chemical weathering. In addition, large quantities of humic acids are provided: the biomass, or weight per unit area, of a tropical forest is two or three times that of a temperate forest. More carbon dioxide is also available. As an example of the rapid rate of weathering, fragments of white mica freshly exposed in a quarry in Madagascar have been transformed into clay within ten years. Hence, one view is that the large amounts of decomposed rocks in the tropics are the result of the intense rate of chemical weathering.

The alternative is that weathering proceeds only very slowly, and the depth of weathering is simply a reflection of the long time that the process has continued undisturbed. In support of this contention, it has recently been demonstrated that the amount of suspended and solute material in tropical rivers is very low. It may be that both views are correct for different localities in the tropics.

Whatever the rate of weathering, one characteristic feature of the process is the formation of hard crusts, or **duricrusts**, near the ground surface. These duricrusts include ferricretes (iron crusts), silicretes (silica crusts), and calcretes (calcareous crusts) in drier conditions. Iron-rich tropical soils capped by ferricretes are alternatively described as **laterites**, although this term has been incorrectly applied to all tropical soils. Duricrusts are formed either as part of local soil weathering *in situ*, or in many other cases by the percolation of mineralised water downslope. Their importance from the landform point of view is that they act as hard beds in the landscape, forming resistant cappings where valley incision has occurred.

Slopes and Streams

Although the movement of material on slopes in the humid tropics is thought to proceed at an insufficient pace to transport all the weathered debris away, it is probably more active than on many temperate hillsides. Rainfall is heavier in the tropics and more intense; the ground flora is less well developed in a tropical rain-forest than in temperate forest, and the humus layer is thinner with the rapid rate of oxidation. It can be inferred that rainsplash impact and surface wash is probably more important on tropical forest floors than in temperate forests. Only finer material will be moved this way, except where water becomes concentrated at the foot of treetrunks.

Coarser material is moved by **landslides**, promoted by a combination of rotted rock and heavy rainfall. The weathered rock continues well below the vegetation root zone, which is thus unable effectively to stabilise the slope. On slopes greater than about 48°, the waste mantle is too thin for mass movement.

Much of the coarse debris moving down the slopes is itself subject to weathering and it appears that only very fine material reaches the streams. This might in part account for one commonly observed feature of tropical rivers, namely their apparent ineffectiveness as erosion agents. Outside the tropics, waterfalls are usually associated with gorges and channel incision, but tropical rivers often fall over sedimentary structures with scarcely any incision, or where crystalline masses occur, the fall merely covers the surface of the resistant core. In either case, the bed of the river is broader at the fall than above or below it; again, a contrast to temperate rivers.

With the lack of pebble-sized material in the load, the river has no tools with which to erode. A second possible reason for the lack of stream erosion is that rivers in the humid tropics are not subject to marked seasonal fluctuations, and do not experience the irregular floods of temperate rivers which accomplish so much erosion.

Landforms

As in the more arid tropical areas, one of the most noticeable features of savanna and rain-forest regions is the presence of inselbergs arising out of more level terrain. Some of these are composed of granite or gneiss because of the relative resistance of these rocks to chemical weathering, and where this is so, they are called **bornhardts**. The precise shape of these depends on the jointing pattern, but many bornhardts are characteristically dome-shaped, with precipitous sides, large areas of bare rock surface, and an absence of talus at their foot. Very large bornhardts are sometimes described as **sugar loaves**, of which those overlooking Rio de Janeiro are the world's most famous examples. The domed shape of these features is related to the large curved sheet joints which develop by the pressure-release mechanism.

Genetically, some bornhardts may be related to the type of inselberg that results from scarp retreat and pediplanation. Many others appear to be the result of selective deep weathering and originate, according to one generally accepted theory, in the following manner.

In every area of rotted rock there are two denudation surfaces: the ground surface, subject to slope movement and stream action, and the **basal weathering surface** or plane, at the contact between the rotted rock and sound rock underneath (Fig. 7.6). In detail, this surface is irregular and complex. The deepest parts occur where the rock is strongly jointed or where the rock minerals are unstable. Elsewhere there are numerous **corestones**, unrotted rounded boulders. In other places, subsurface domes exist, very similar in shape to exposed bornhardts and strongly suggestive of a connection with them in origin. Thus the first stage in the formation of a bornhardt is selective subsurface weathering which rots the rock around it. The second stage is its exposure by removal of the rotted rock. This may occur over a long period of time by normal stream activity, or be accelerated by base-level

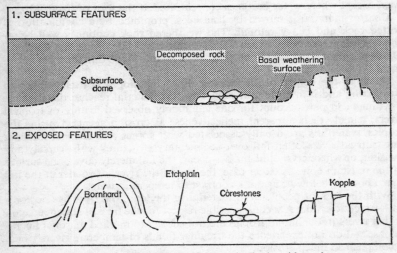

Fig. 7.6. Deep weathering features in the humid tropics.

alterations or climatic change. Once revealed, the form of bornhardts may be accentuated because they shed water off themselves and rapidly translate it to the surrounding rotted rock.

When the basal weathering surface is revealed, it is called an **etchplain** (Fig. 7.6), so called because of the way that its features have been etched out in the first stage of subsurface rotting. In addition to the dome forms, other more rectangular inselbergs, **koppies**, will be revealed, as well as individual corestones. This hypothesis of rotting and stripping has been applied to the tors of Britain. The deep rotting episode took place at a time when Britain experienced a climate warmer than now.

In summary, then, we can see that in the tropics similar landscapes can be produced by different processes, by pediplanation or by those in the etchplain model. In savanna lands, areas intermediate between the arid and humid extremes, both processes may well be operative together, or have alternated with successive climatic changes. But it must be admitted that our knowledge of these landscapes is still rather limited, and much must now rest with general theory.

Suggested Further Reading

Bagnold, R. A., *The Physics of Blown Sand and Desert Dunes*, Methuen, London, 1941.

Cooke, R. U., and Warren, A., *Geomorphology in Deserts*, Batsford, London, 1973.

Cooke, R. U., 'Deserts', in *The Unquiet Landscape*, edited by Brunsden, D., and Doornkamp, J. C., IPC Magazines, London, 1974.

King, L. C., *Morphology of the Earth* (2nd edn.), Oliver & Boyd, Edinburgh, 1967.

Thomas, M., 'Landforms in Savanna Areas', in *The Unquiet Landscape*, op. cit.

Tricart, J., *The Landforms of the Humid Tropics, Forests and Savannas*, Longman, London, 1972.

CHAPTER EIGHT

COASTAL FEATURES

The coastline provides the geomorphologist with a unique range of different environments. On the one hand, wind and wave forces vary markedly from place to place, both locally and on a global scale. High-energy conditions prevail on the exposed west-facing coasts of temperate latitudes, and these contrast with the quieter conditions of many tropical shorelines. Equally, the character of the coastline itself varies considerably and this produces a wide range in the rate of morphological response to marine processes. **Sandy or pebbly beaches** rapidly adjust to the daily variations in wave conditions. Here the concept of dynamic equilibrium can be applied to good effect, for beach profiles, spits and similar depositional features often represent in their form a fine condition of balance with the controlling factors.

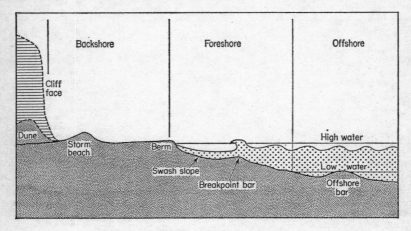

Fig. 8.1. The terminology of coastal zones.

Cliffed coastlines have a slower rate of response. Although some cliffs can be seen to be eroded over the course of a year or so, many others appear to change hardly at all, and contain in their form the legacy of long past events, such as different sea-levels. These variations make coasts both interesting and complex.

Some of the principal elements which make up a typical coastal area are illustrated in Fig. 8.1. What is popularly termed 'the shore' includes both the **foreshore**, which stretches from the lowest tide limit to the mean high tide limit, and the **backshore**, which continues the shore to the extreme limit of high tides and storm waves. The term **beach** is used in this chapter to define an accumulation of marine-deposited pebbles, sand or silt.

Factors Affecting Coasts

We can usefully outline here some of the background factors which need to be borne in mind when considering the different types of coast later in the chapter. Of course, coastal features are primarily shaped by the processes working on them, but these include not only wave erosion, transport and deposition, but also subaerial processes such as weathering and mass movement on cliffs. These factors apply to lake shorelines as well as sea coasts.

The marine agents of erosion and deposition, waves, tides and currents, are controlled by **oceanographic** and **climatic** factors. Particularly important here are the Earth's atmospheric circulation features (Chapter Thirteen), which control wind direction and force, and hence the generation of waves and currents. Climate affects the rate of weathering on cliffs and also partly controls the biotic factor in coastal environments. **Geological** factors of structure and lithology influence the general trend of the coastline, its detailed configuration, cliff profiles and offshore gradients. Rock type also controls the local type of beach sediment. The **biotic** factors to be considered include that of vegetation, which plays an important role in both sand-dune areas and also in low-energy environments such as salt-marshes and mangrove-swamps. In addition, many tropical coastlines are distinctive because of the presence of coral and other reef-building organisms. **Man** is also a fundamental factor in coastal geomorphology, and has affected coastal processes through reclamation projects, coastal protection works, leisure activities and pollution.

The factor of **sea-level change** is one of the most important in coastal geomorphology. Some illustrations will be given later in the chapter of its role in the origin of certain depositional and erosional landforms, which are otherwise difficult to explain solely in terms of modern processes. Sea-level changes occur as a result of (*a*) **eustatic** or global changes in the level of the sea, as has occurred with the successive growth and wane of ice-sheets during the Pleistocene, or (*b*) **tectonic** changes in the level of the land. These tectonic changes may be **isostatic**—that is, adjustment to a load, as in glacial isostasy (Chapter Six)—**epeirogenic**, related to broad scale tilting; or **orogenic**, related to folding and flexuring. In Britain, glacial eustasy and glacial isostasy account for many of the sea-level changes that have affected the coast.

The level of the sea has been both higher and lower than now, and has left **emergent** (raised) and **submergent** (drowned) features, sometimes on the same stretch of coastline. Submerged effects include **rias** (drowned river valleys), **fjords** and submerged forests. **Raised beaches** and **platforms** and abandoned clifflines are the main emergent features. An outline of the sequence of Pleistocene sea-level changes is sketched in the next chapter.

In areas affected by isostatic recovery (Chapter Six), a complex series of raised beaches often exists, reflecting the interplay of isostatic and eustatic causes. One of the highest raised beaches in Britain is found at Stirling at +37 m above ordnance datum (O.D.). This was formed some 13,000 years ago, when the world-wide sea-level lay at approximately −50 m O.D. For the beach to be at its present elevation above sea-level, there has therefore been 87 m of isostatic recovery at Stirling since 13,000 years before present (B.P.). When traced for many kilometres, all the Scottish raised beaches become progressively lower away from the centre of glaciation in the south-west

Grampians. It is therefore misleading to refer to an isostatic beach by a single height, as they are in fact all tilted.

Sea-level change is still taking place. On the evidence of tide gauge records, world-wide sea-level has been rising for the past fifty years at an average of 1·2 mm per year. Faster rates of land/sea-level change occur in areas undergoing crustal warping. In northwest Europe, the Baltic area is still rising at rates up to 10 mm a year (Fig. 8.2).

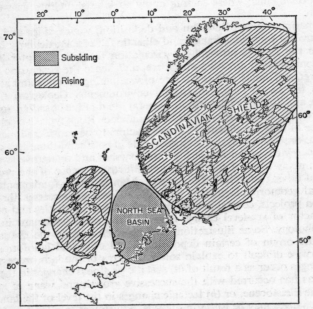

Fig. 8.2. Contemporary sea-level change in north-west Europe, based on tide gauge measurements (figures in mm/yr).

Shoreline Processes

Waves are by far the most important agents of shoreline modification. Apart from those very occasionally started by earthquakes, all other waves are produced by wind. The wind exerts a drag on the surface-water particles and sets up small orbital motions in the water, largest near the surface, becoming less with depth (Fig. 8.3). There is some small displacement of the water, but the wave itself moves much faster through the sea in the direction of the wind, growing as it travels. Three factors govern the size of the wave: the wind speed, the duration of the wind, and the distance or 'fetch' over which the wave travels. The largest waves will inevitably be produced by prolonged gale-force winds blowing over ocean surfaces. Waves reaching a coast after travelling long distances form a **swell**; those generated by local winds create a **sea**.

When waves reach shallow water near a coast, the sea floor alters the dimensions of the wave (Fig. 8.3). Although the wave period of time (T) remains the same, the wave slows down, such that the wavelength (L) is shortened, but the wave height (h) increases. The wave steepness (h/L ratio) increases until it becomes so steep that it breaks. In the breaker or surf zone, the wave force is translated up the beach and creates a **swash**. This returns as **backwash** either in a sheet flow (undertow) or as a **rip-current**, a localised concentration of backwash, which can be very dangerous to bathers. **Wave steepness** is a critical parameter in making waves either constructive or destructive. **Constructive waves** (spilling breakers) have a low index of steepness; **destructive waves** (plunging breakers) have a relatively high h/L ratio. We shall see the workings of this on beach profiles later in this chapter.

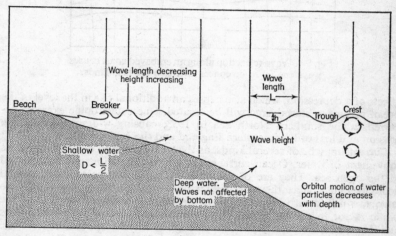

Fig. 8.3. Wave form in deep and shallow water.

Waves are also subject to **refraction** as they approach the coast. Where oblique waves approach a straight shore, the frictional drag exerted by the sea floor turns the waves to break more nearly parallel to the shore. On an embayed coast (Fig. 8.4), headlands interfere with the waves first and set up refraction such that wave attack is concentrated on the headland and is much reduced in the bays.

Tides are movements of the ocean set up by the gravitational attraction of the sun and moon. They are predictable in timing and level, varying from high spring tides, when earth, moon and sun are in alignment, to low neap tides, when lunar and solar effects are opposed. The most important effect of tides is that their range controls the vertical range of wave action. A large tidal range (macrotidal environment) creates a broad shore zone; a small tidal range (microtidal environment) concentrates wave energy at a more constant level. Tidal range can vary locally according to the configuration of the coast: the highest tides in the world occur in estuaries, such as the Bristol Channel and the Bay of Fundy, where the maximum range is more than 15 m. Onshore winds can locally increase the height of a tide; where they combine with the

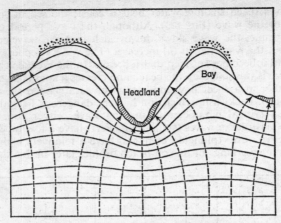

Fig. 8.4. Wave refraction along an embayed coast causes
wave energy to be concentrated on the headlands.

effects of a depression, which also causes an additional rise in the level of the
ocean surface, a **storm surge** will be created. One catastrophic example occurred in the southern North Sea in 1953, causing widespread flooding,
erosion and loss of life in eastern England and the Netherlands.

Currents may be of several kinds, as the term is used to describe any mass
movement of water. Ocean currents are major circulations set up by wind
in the open sea. They are of considerable indirect importance to coastal
geomorphology, since they act as a major control on the supply of offshore
energy, which in turn influences wave dimensions, sediment supply, and the
biotic factor in coastal areas. Tidal **currents** are created as the tide ebbs
and flows, and are usually powerful enough to move sediments in estuaries.
The **ebb tide** is usually the stronger, as it is reinforced by the outflow of river
water. Local winds may create currents in gulfs and estuaries and locally along
the shore. Recent evaluations suggest that the role of currents in coastal
studies has for long been underestimated, but they are probably more
effective in shaping sea-floor topography than modifying beaches. Contrary
to earlier beliefs, the longshore movement of beach material is no longer
attributed to currents, but to wave action.

Cliffs and Platforms

Active sea cliffs occur on exposed coasts where the waves are able to reach
the foot of the cliff and are not prevented from doing so by large accumulations of sand or shingle. Paradoxically, the presence of a limited amount of
beach material can aid the sea to erode the cliff. Except where the offshore
water is very deep, most cliffs are fronted by a wave-cut platform which
develops as the cliff retreats.

Cliff Erosion and Morphology

The marine erosion processes at work at the foot of a cliffed coast are of
three main types. **Physical abrasion** and corrasion are important where beach

material is available. Some of this material may be supplied by longshore drift, or from the cliff itself. The sea uses the material as tools of erosion much in the same way as a river. Both the material and the foot of the cliff become rounded and smoothed in the abrasion zone. A certain amount of **chemical erosion** may take place: sea water is very effective in dissolving a number of rock-forming minerals, especially hornblende, orthoclase and carbonates. Chemical action will be important in both the wave and the spray zone. **Hydraulic action**, the force of the waves themselves, becomes effective where it produces tremendous instantaneous compressions of air in cracks and crevices, suddenly released as the wave recedes. On suitably jointed strata this has the effect of weakening the structure and quarrying blocks of rock.

In addition, sub-aerial slope processes (Chapter Four) are very important on sea cliffs. Some of the largest landslides and rockfalls are marine-induced by undercutting. In less exposed situations, weathering and mass movement may produce an **undercliff**; the role of the sea is relegated to one of removing the products of sub-aerial erosion.

The **shape** of cliffs is controlled by the lithology and structure of the rocks and the relative effectiveness of marine as opposed to sub-aerial processes. We must also take into account the past geomorphological history of the cliff. Competent rocks tend to give rise to vertical cliffs; so do those that are horizontally or vertically bedded, or have landward-dipping strata. Those with seaward-dipping strata more usually recline at a lower angle. On hard rocks both major and minor weaknesses in the rock structure have a detailed effect on the plan and profile of cliffs. Joints and faults may be enlarged into deep narrow inlets (**geos**) and even into caves, tunnels or blowholes. Where hard rocks overlie soft, this arrangement is often conducive to the formation of large-scale landslips, as at Folkestone Warren and Seaton in Devon. The alternation of hard and soft lithologies in plan view results in features of **differential erosion**, creating bays and headlands, of which examples abound around the British coast. Cliff erosion will be most rapid on completely uncemented rocks. In Britain, these include the Pleistocene deposits of the east coast and the Tertiary rocks of the London and Hampshire basins. Losses averaging up to 3 m a year have been recorded at some localities. Hand in hand with erosion rates as rapid as these, effective longshore drifting or removal offshore must also take place to prevent the eroded material accumulating at the foot of the cliff.

Within the limits set by lithology and structure, the largest active cliffs occur where the exposure to waves is greatest. Cliffs are nearly always most vertical around headlands rather than within bays. The largest cliffs in Europe occur in Portugal and on the west coast of Ireland, facing the biggest Atlantic swell waves. Where, through a combination of rock type and exposure, the rate of retreat is especially rapid, **hanging valleys** may be left, well exemplified by the Seven Sisters cliffs in Sussex.

Finally, as an example of the inheritance factor on cliffs, many hard rock cliffs are **bevelled** in their upper parts. In some cases, the cause may be structural or related to present sub-aerial process. But bevels are often associated with fossil scree deposits resting on raised platforms at the foot of the cliff, and hence cliff bevels in these situations appear to be the result of past periglacial action.

Platforms

Shore platforms develop in the intertidal zone at the foot of cliffs. They are generally attributed to marine abrasion, and have a gently concave slope to seaward, which ideally is an equilibrium profile just steep enough to allow debris to be moved by wave action (Fig. 8.5). The junction between platform and cliff is marked by a smooth rounded **notch** which lies just below high-tide level. The seaward limit of the abrasion platform is determined by the maximum depth at which material can be moved by waves. This is about 10 m below low tide level. Thus the overall height range, and hence the width of the platform, vary according to the tidal range and wave height. Platform notches will be slightly higher around headlands than in sheltered bays. Given optimum conditions, the maximum width of an abrasion platform will be of the order of 500 m. Much wider platforms than this have been suggested to explain ancient erosion surfaces (Chapter Nine). The only circumstances where a very wide platform could be created would be on a steadily rising sea-level.

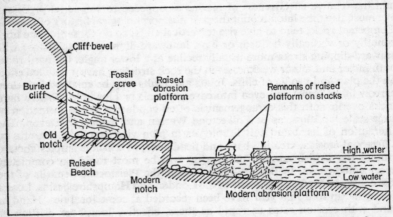

Fig. 8.5. Morphology of a cliffed coast with fossil and modern abrasion platforms.

It has been suggested that processes other than abrasion might create platforms. One possibility is the concentrated action of chemical and biological agents in the intertidal zone. Around British coasts, limestone outcrops are characteristically pitted and honeycombed, showing that abrasion is not the only process at work. In tropical waters, chemical and biological activity are major factors in promoting extensive **reef-flats** in the intertidal zone. These flats undercut the adjacent cliff forming an overhanging **visor**.

Another possibility is the process of **water-layer weathering**, which has been recognised around the coasts of Australia and in similar subtropical latitudes. Platforms are created by this process at or above the high-water mark in the spray zone, where alternate wetting and drying of the rock leads to its gradual disintegration. A bench is formed as storm waves sweep away the weathered debris. However, it has been suggested that these platforms may at least in part be old abrasion platforms formed at a higher sea-level.

Relict or **raised platforms** have been widely recognised as evidence of former high sea-levels. Good well-planated examples are widespread in western Britain and Ireland; their marine origin may be confirmed by overlying beach deposits (see Fig. 8.5). It is also likely that some of the platforms now being washed by the present sea-level may be exhumed features. Modern platforms are not usually well planated, but where they are, the alternative explanations are either that the rock is easily erodable, as in the case of Chalk platforms, or that the platform was cut long ago and is simply being retrimmed by the modern waves.

Beaches

Beaches are usually composed of either sand or shingle. On a global scale, sandy beaches predominate; shingle beaches are more common in higher latitudes, where there is more coarse material supplied by present and past glacial and periglacial action. In southern and eastern England, much of the shingle is flint, derived from Chalk cliffs. Beach sediments also tend to be well rounded and well sorted, in response to the constant energy exchange on the shoreline. Waves move beach material both up and down the beach and also along the coast.

Beach Profiles

Some beaches are steep, others very flat; for many, their cross-profiles are different in winter and summer. Shingle beaches are normally steeper than sandy beaches, but within the limits set by the size of material, the main controller of beach profiles is wave steepness (h/L). This is because the amount of material piled up or moved down a beach varies with the type of wave. We can identify the relationships in a simple model:

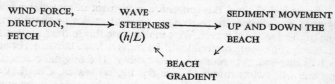

Steep waves comb down the beach by moving sediment seaward in their powerful backwash. However, in reducing the beach gradient, this allows the wavelength (L) to become longer, automatically reducing the wave steepness. Thus the process of combing down does not go on indefinitely, and an equilibrium is reached between the beach gradient and the prevailing waves. This is an illustration of **negative feedback** between form and process. Similarly, flat waves shift sediment landward, steepening the beach; but this increase in gradient will eventually shorten the wavelength, making the waves more destructive. Again a balance is achieved.

In summer, when constructive waves predominate, a **berm** is built up at the limit of wave action. Sometimes several berms may be observed, related to a spring-neap tide sequence. In winter, storms produce much flatter profiles, especially on sandy beaches. However, on a shingle beach, a **storm beach** may be thrown up in the form of a ridge on the backshore. It remains there because the coarseness of the shingle allows the backwash to percolate through it rather than wash it away.

Beach Orientation

Material is moved along beaches by waves breaking obliquely on the shore. The sand or shingle is pushed up the beach at an angle by the wave, but when it returns in the backwash, it does so down the steepest gradient, normal to the shoreline. Each particle therefore moves in a zigzag fashion (Fig. 8.6).

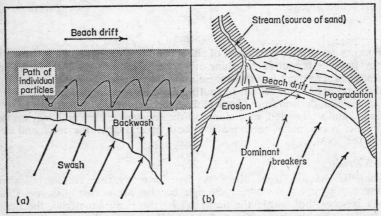

Fig. 8.6. (a) Mechanism of longshore drifting. (b) Reworking of sand by longshore drifting aligns bayhead beach at right angles to the dominant breakers.

Over a long period of time, this longshore movement of material means that beaches attempt to orientate themselves at right angles to the direction of approach of the dominant waves. We can illustrate this in the simple case of a **bayhead beach** (Fig. 8.6): material will be moved from left to right eroding the beach at one end, prograding it at the other. The original orientation of the shoreline will swing round to face the biggest waves. Chesil Beach in Dorset is a structure almost perfectly aligned at right angles to the approach of the main Atlantic waves.

This theory, first suggested by W. V. Lewis, and known as the **dominant breaker hypothesis,** is an important principle underlying the evolution of many beaches, spits and bars. However, equilibrium orientation is not always clearly evident, and it is not always so well seen in sand as in shingle structures. Steep offshore water, strong currents and local refraction in bays may all mask the relationship between the beach and the dominant breakers.

Beach Structures

Spits are a very common form of depositional structure, where bays and headlands exist, and are usually sited at points where the shoreline undergoes a sharp change of direction, as at the entrance of a bay or estuary. Longshore movement of material extends the beach into open water. Spits are usually

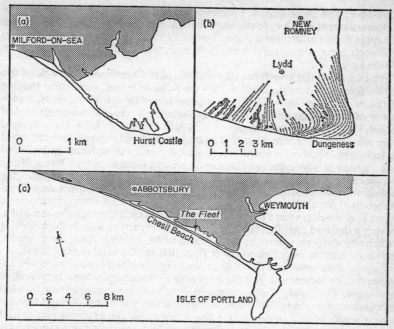

Fig. 8.7. Beach structures: (a) spit with recurves; (b) cuspate foreland; (c) barrier beach.

recurved at their extremities (Fig. 8.7a), the recurves being produced by wave refraction round the head of the spit, or by wave approach from a different direction. As the spit extends, several recurves may be built. Where longshore drift creates a spit connecting an island to the mainland, this forms a **tombola**. Drift of material along a shore from opposing directions results in the building of a pointed depositional structure known as a **cuspate bar** or **foreland**. Dungeness, in Kent, is a classic example of a cuspate foreland formed by two sets of dominant breakers which have built up a series of shingle storm ridges over the period of many centuries (Fig. 8.7b).

Offshore bars can often be seen on gently shelving coastlines and where the bars are long and continuous they form **barrier beaches**. Some bars may originally have been created by longshore drifting—in other words, they are extended spits. The large barrier beaches, as exemplified by those on the east coast of the United States, appear to have been piled up parallel to the shore by the post-glacial rise of sea-level. Several of these structures appear still to be slowly migrating landwards and may eventually become part of the bayhead beach. In England, Chesil Beach is a shingle barrier structure of similar origin (see Fig. 8.7c).

Coastal Dunes

Active coastal dunes occur only where there are sandy beaches and an adequate supply of sand in the coastal transport system, and where the wind is onshore. The biggest coastal dunes are found on west-facing coasts in the temperate storm belts of the world. The main difference between desert dunes (Chapter Eight) and coastal dunes lies in the presence of vegetation, mainly sea-couch grass and marram grass in Britain, which has a trapping effect on blowing sand.

We can recognise two types of dune in coastal areas. At the back of the beach, **foredunes** develop in the form of a ridge or ridges, parallel to the line of the beach. A dried-out berm may provide the initial foundations. Landwards of the foredunes, there occurs a seemingly chaotic accumulation of **secondary** or **transgressive dunes**, which will originally have been derived from erosion of the foredunes. This secondary accumulation of sand can build up to heights of 30 m or more. Recognisable regular dune forms to be found here include **parabolic dunes** of crescentic ground form with their horns facing the wind. This is opposite to barchan form because vegetation impedes the migration downwind of the lower slopes of the dune. A **hairpin dune** is an exaggerated form of parabolic dune. **Blowouts** are quite common in sand-dune areas, and develop where the wind manages to breach through the dune-line, leaving a deflated area flanked by lines of sand on either side. If the hollow is sufficiently deep to reach the local water-table, a damp **dune slack** results, attracting a more varied plant cover than that on the mobile sand areas.

Coastal dune areas are liable to change rapidly if the vegetation cover becomes thin or is removed, and the sand may encroach on nearby farmland or settlement. The fertile Culbin estate in Scotland was overwhelmed by sand encroachment at the end of the seventeenth century and transformed into the barren waste of Culbin Sands. On the other hand, the artificial planting of marram grass or trees can stabilise dunes, and this has been successfully employed at Culbin.

Low-energy Coasts

In contrast to the high-energy conditions experienced by many beaches, sand-dune areas and exposed cliffs, much quieter conditions exist within estuaries, behind spits and bars and in deep embayments. Here finer sediments settle out forming **mud flats**, allowing salt-loving plants to colonise much of the area.

In such environments, three zones related to tidal activity may be recognised (Fig. 8.8). The **sub-tidal zone** is made up of permanently occupied creeks and channels. These channels are often intricately meandering and exhibit many of the features of a dendritic drainage system. The **intertidal slopes**, covered by every high tide, compose the muddy flanks of the channels. These are normally bare of vegetation, except near their upper parts. Just above the mean high water mark, the **high-tide flat** is, as its name suggests, a level surface built up at the extreme limit of deposition.

These three units can be recognised in a wide variety of climatic and vegetational environments. In temperate areas, the high-tide flat is colonised by **salt marsh**, usually dominated by cord grass (*Spartina*) and marsh samphire (*Salicornia*). In tropical areas the high-tide flat is characterised by **mangrove swamps**.

There has always been some debate as to the precise role of salt-marsh plants and mangroves in trapping sediment and helping to build the mud flats. There is no doubt that once established, they do aid deposition and also

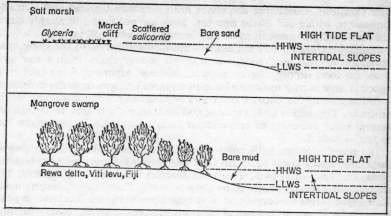

Fig. 8.8. Low-energy coasts: temperate (*top*); tropical (*bottom*).
(After J. L. Davies, courtesy Longman)

have important effects on minor relief forms, such as the channel systems. However, the modern concensus is that the vegetation is secondary, and is essentially taking advantage of locations where sedimentation is occurring in any event because of the physical conditions. Marshes in low-energy zones are very much dependent on protection, and if this is removed for any reason, such as with the migration of a spit or bar, then erosion is usually rapid.

Coral Coasts

Coral reefs are uniquely different from any of the coasts previously discussed, in that they are composed entirely of matter accumulated through organic processes. The reef is built by the corals secreting lime in forming

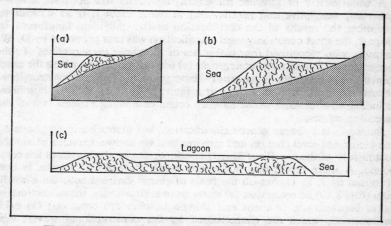

Fig. 8.9. Coral reefs: (a) fringing reef; (b) barrier reef; (c) atoll.

their skeletons. Coral reef shorelines are essentially tropical and virtually confined to within 30° of the equator. They are not found in muddy water and corals will not grow in water any deeper than about 50 m.

Fringing reefs (Fig. 8.9) are built as platforms attached to the shore, and are widest around headlands, where corals receive clean, fresh water with abundant food supply. **Barrier reefs** lie offshore, separated from land by a lagoon. Large barrier reefs may be some distance offshore from the mainland, as in the case of the Great Barrier reef adjacent to the Queensland coast in Australia. This reef is 2000 km long and 1000 m wide, broken by occasional passes. **Atolls** are more or less circular reefs enclosing a lagoon which has no land inside it.

The main problem with coral coasts is to account for their origin; in particular, many coral reefs are built in water of depths well beyond the theoretical limit of growth. Two main theories have been put forward. The **glacial control theory**, suggested by Daly, states that coral shorelines originated during glacial low sea-levels, and have built up with the sea-level rise, in post-glacial time. However, borings of many atolls have shown that coral structures exist to depths up to 760 m, well beyond the lowest glacial sea-levels. The alternative **subsidence theory** was first proposed by Darwin in 1842, and still receives a wide measure of support. He supposed that small islands have slowly subsided with the general downwarping of oceanic parts of the Earth's crust. Thus, what were originally fringing reefs became barrier reefs and eventually atolls as the land core submerged. With different rates of subsidence, we now see coral shorelines at all these stages of development. Recent geophysical evidence shows coral bases at a wide variety of depths, which suggests that the subsidence theory is the more feasible in explaining the deepest coral formations. However, practically all reefs record evidence of eustatic oscillations in the form of submerged notches and benches, and hence glacial control may well account for some of their detailed morphological features.

Coastal Classification

A wide variety of schemes for classifying coasts has been suggested—**numerical, descriptive** and **genetic**—but in many cases it is as difficult to remember the results of the classification as the individual landforms. In the past, the most commonly used classification was that proposed by D. W. Johnson, who recognised four categories of shoreline: (*a*) shorelines of submergence; (*b*) shorelines of emergence; (*c*) neutral shorelines, where the exact form is due to neither emergence nor submergence, but to a new constructional or tectonic form, such as a delta or a fault; and (*d*) compound shorelines, including all shorelines which have an origin combining at least two of the preceding classes.

Johnson's is a simple genetic classification, but places heavy emphasis on one factor, sea-level change, and requires that we know something about the past history of the coastline. Strictly speaking, nearly all shorelines are compound. A more recent and potentially much more useful scheme is that proposed by J. L. Davies on the basis of **energy environments**. On a world map (Fig. 8.10) he recognises (*a*) **storm wave environments**, where destructive storm breakers are frequent and shingle beaches are common; (*b*) **swell environments**, which are characterised by flat constructional waves; and (*c*) **protected environments**, the low-energy conditions of enclosed or partially

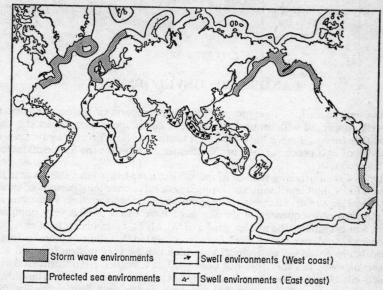

Storm wave environments   → Swell environments (West coast)

Protected sea environments   ⌃ Swell environments (East coast)

Fig. 8.10. A classification of coastlines according to energy characteristics.
(After J. L. Davies, courtesy Longman)

enclosed seas. It is also possible to apply the concept of differing energy environments on a more local scale, as has been illustrated with some of the features described in this chapter.

Suggested Further Reading

Bird, E. C. F., *Coasts*, Australian National University Press, Canberra, 1968.

Davies, J. L., *Geographical Variation in Coastal Development*, Oliver & Boyd, Edinburgh, 1972.

Fairbridge, R. W., 'The changing level of the sea', *Scientific American*, **202**, 1960.

King, C. A. M., *Beaches and Coasts* (2nd edn.), Arnold, London, 1972.

Zenkovich, V. P., *Processes of Coastal Development*, Oliver & Boyd, Edinburgh, 1967.

CHAPTER NINE

LANDSCAPE DEVELOPMENT

Our consideration of geomorphology in this section of the book has been largely concerned with an understanding of contemporary processes and the resulting forms. But we have also seen that some landforms may retain the imprint of past processes, and in this chapter we examine the historical factor in more detail.

One of the intriguing themes in the subject has always been the manner in which individual landforms and landscapes evolve over long periods of time. At a teaching level, attempts to integrate the seemingly endless variations in landforms into sequences of development have often proved very popular. This was essentially the master stroke of W. M. Davis in his cycle of erosion. However, modern research work has demonstrated that some of these teaching models are based on generalisations which can be very misleading, and the Davisian cycle has therefore not been used as a basis for landform study in this book. However, we should certainly not reject time as a factor in landform studies any more than we should accept it as the only framework for the subject; rather in our present state of knowledge we can evaluate its role.

Reconstructing Landscape Development

Two broad approaches can be employed in reconstructing the way land-forms may have evolved. On the one hand, an inductive approach can be adopted, as employed, for example, in the postulation of long-term cycles of erosion. We attempt to induce evolutionary models by applying what we know about the variable shape of landforms and by using our knowledge of processes and the rates of change in landforms.

In the alternative approach we try to deduce what has happened by recognising and dating specific events in the evolution of the landscape. This involves the study of deposits or erosional facets, and the adoption of strati-graphic principles and techniques. This kind of approach has been widely used in reconstructing the evolution of the landscape in our present epoch, the Pleistocene. Documentary evidence may also be useful in dating events in historical times. For the very recent episodes in landscape development, such as the record of flooding, data are often sufficiently precise for geo-morphologists to be able to pay special attention to the relationship between the magnitude and frequency of events and the role of each in changing landforms. Empirical data of actual events can be used in this way to predict future evolution.

Cyclic and Non-cyclic Development

One of the most persuasive ideas in geomorphology has been that land-scapes develop in cycles. Suggestions have been made for various erosion cycles in different types of geomorphological environment, but the underlying theme in all cases is that each starts from an **initial state** which passes through a sequence of events to end in an **ultimate state** approximating to the original surface form. A new cycle is initiated with a change in base-level, usually caused by uplift of the land, which injects new energy into the erosional system.

The Davisian Geographical Cycle

The first well-ordered example of a cycle of erosion was put forward in 1899 by W. M. Davis, and called by him the Geographical Cycle. This cycle dominated geomorphological thought for the next fifty years. Davis pointed out that there were three main controls on landscape development: **structure**, meaning all rock characteristics; **process**, the mechanisms of landscape change (Chapters Three to Eight); and **stage**, the length of time the processes have been operating. Largely concentrating on the last factor, Davis divided the cycle into the relative stages of youth, maturity and old age.

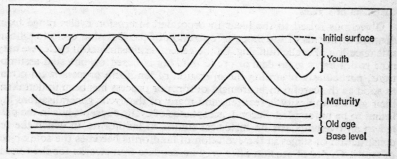

Fig. 9.1. Stages in a cycle of erosion (according to Davis).

To demonstrate his cycle in the simplest way, Davis assumed that it was initiated by the relatively rapid uplift of a block of land, which ideally would be flat, but may be any other geological structure. In the stage of **youth** (Fig. 9.1), streams begin to trench steep-sided V-shape valleys into the initial landmass, increasing the available relief. Progressively through youth the landscape becomes dominated by valley slopes. The landscape passes into **maturity** when the valley sides intersect at the interfluves and the last traces of the initial landsurface disappear (Fig. 9.1). At this stage the rivers begin to form flood-plains and the rate of downcutting in the valley floors drops markedly. Through maturity, the relief of the region decreases steadily and slopes become progressively lower in angle. The stage of **old age** (Fig. 9.1) is reached when the landscape is reduced to a low undulating surface in which streams have very low gradients and extensive flood-plains. Davis gave the word **peneplain** to this surface of low relief. Interruptions to the cycle might be caused by climate or base-level change, leading to the **rejuvenation** of the system. A landscape might therefore become **polycyclic**, carrying the traces of two or more partially completed cycles.

Other Cycles of Erosion

Davis originally intended that his cycle should be the normal one for all landscapes. He later considered that landscapes in glacial and arid landscapes, referred to by him as **climatic accidents**, were sufficiently unique to warrant separate cycles of glaciation and desert erosion.

Other geomorphologists have suggested modifications to the Davisian cycle, or proposed wholly new ones for various climatic and structural environments. Of the more notable, D. W. Johnson, proposed a cycle of shoreline development; the Yugoslavian geologist J. Cvijic suggested a cycle of karst erosion; a cycle of periglacial erosion was drawn up by L. C. Peltier; and a savanna cycle was put forward by C. A. Cotton. (For details of these schemes, readers are referred to the suggested reading at the end of the chapter.) Lester King's pediplanation concept (Chapter Eight) was also put forward by him in a cyclic context. King has proposed that the pediplanation cycle, proceeding by the parallel retreat of slopes and leaving nearly horizontal pediplains as an end-product, is in fact the normal one for the world, rather than the Davisian cycle with its peneplain.

Criticism of Cycles

Objections raised to the ideas incorporated in erosion cycles range from detailed criticisms of the mode of operation of the Davisian cycle to wholesale rejection of cyclic thinking. Of the general criticisms about cycles we may note two. First, a great deal of cyclic thinking is based on untested assumptions, particularly regarding the operation of landform processes. So often, as soon as the careful measurement of form or process has been undertaken, their great complexity is realised, and many of the cyclic generalisations are found to be unfounded. Second, acceptance of cyclic concepts has in the past led to an overemphasis on historical studies in landforms, in which the reconstruction of stages in the evolution of landforms becomes the sole object of study. Cycles of erosion offer very inadequate frameworks for the fundamental study of process/form relationships.

Turning to the Davisian cycle itself, most geomorphologists now regard his description of the progression of the cycle through its various stages as inaccurate if not misleading. There is little evidence that landforms do evolve, as he described it, to an inevitable end-result, the peneplain. In practice there are usually no grounds for the Davisian assumption that steep slopes are 'young' and gentle slopes are 'old'. In Chapter Four we stressed some of the many variables that control slope angle. In other words, there is the fundamental objection to the idea inherent in the Davisian cycle that the angles of slopes or the channel patterns on a river inevitably tell us something about their age. We may also doubt whether flat surfaces can ever be created entirely by slope decline, as the energy for work in the landscape becomes impossibly low in the later stages. There is also the further point that the time for the completion of a Davisian erosion cycle is perhaps to be measured in many millions of years, and one may wonder whether such a cycle could ever be completed in view of the known rapidity of base-level changes in the later Cainozoic era. Although there are many low-relief surfaces in the world today, it is difficult to identify one positively as a peneplain as Davis described.

Non-cyclic Development

We have seen in our review of landforms processes in earlier chapters that the landscape can be usefully viewed as an open system, in which the variables in the system attempt to regulate themselves towards the establishment of equilibrium states. Taking this a stage further, several geomorphologists, notably the American, J. T. Hack, have suggested that once a condition of dynamic equilibrium is obtained between landforms and the environment, there need be no further changes to the shape of landforms. This does not mean that mass and energy do not continue to flow through the system, but the individual morphometric properties of the landscape become independent of the reduction of relief and the passage of time. Only when the environmental factors change, such as with climatic change or with variation in the rate of uplift, will there be readjustment of form.

This non-cyclic view of the landscape is essentially an attempt to consider landform evolution within fairly narrow limits. If we take the total landscape over a long period of time, there must inevitably be a tendency for a gradual reduction of mass in the system. As geomorphologists R. J. Chorley and B. A. Kennedy have pointed out, the position is basically that there are constant adjustments made in every landscape to new steady-state conditions, but these are superimposed on a general tendency for change associated with the reduction of relief through time. The result is landscape evolution, but we can stress again, this reduction in relief is not necessarily accompanied by an inevitable change in all aspects of the geometry of the landscape, as suggested by the Davisian cycle. Nor is it inevitable that long-term evolution should end in every case with a situation of low relief, whatever its origin, to be followed by a fresh start.

Time is therefore both important and unimportant in geomorphology, depending on the scale and time-span of what one is studying. The working of individual processes in the present-day landscape is best studied on a time-independent basis; on the other hand, the form of many landscapes inevitably reflects timebound aspects, related to events over a long period of time. Timeless and timebound approaches are both necessary to the full understanding of any landscape. These approaches are not so incompatible as some writers have made out. Both long-term change, which may be cyclic, and steady states can be embraced in the concept of the landscape as an open system. This, rather than a cyclic framework, offers the most flexible approach to landform study. Cyclic thinking offers one perpective on geomorphology, but is inadequate by itself for the present-day study of the subject.

Planation Surfaces as Indicators of Landscape Development

Planation surfaces, sometimes called erosion surfaces, have been widely used in geomorphology as indicators of the various stages in the evolution of individual landscapes. The term is applied to areas of land which are flat or have a very low relief, and which are believed to be the end-product of long periods of erosion. There are several possibilities as to the processes which actually created them: planation surfaces may, for example, be pediplains, etchplains, altiplanation surfaces, marine-cut platforms, or for those who follow Davisian ideas, peneplains.

Although planation surfaces were recognised in the nineteenth century, the Davisian concept of a peneplain gave the main impetus to the idea that flat surfaces could be used as indicators of past base-levels. From this, the recognition of planation surface remnants at different heights in the landscape led to attempts to reconstruct the **denudation chronology** or history of erosion of many areas in the British Isles and elsewhere. But it is now recognised that the concept has been too uncritically applied on several counts; in particular low-relief surfaces or benches do not have to be the end-product of vast periods of erosion, but can be locally created at all stages, as in pediplanation.

Great difficulties arise with both the recognition and interpretation of planation surfaces. For instance, **accordant summit heights** on hilltops have sometimes in the past been regarded as representing the last vestiges of an ancient planation surface, on the theory that the oldest relief should be preserved on interfluves. But accordant summit heights may also be produced simply by the intersection of valley sides in adjacent equally spaced river valleys. Even if planation surfaces can be positively identified, their mode of origin presents further problems. In Britain, many surfaces used to be regarded as peneplains, but it is accepted that other modes of origin, for example pediplanation, may be more likely.

The most successful reconstructions of landscape evolution using planation surfaces have usually relied on supporting evidence, particularly the presence of superficial deposits such as marine gravels, or duricrusts. These are often much more reliable guides to the origin and age of the surface than the form itself.

Planation Surfaces in Britain

In marked contrast to the extensive planation surfaces of areas such as southern Africa, the planation surface evidence in Britain is puny, and has fostered a host of different interpretations. In one of the more recent syntheses of the available evidence, D. L. Linton has commented on the broad similarity of form in each of the British highland areas, despite their different elevations. A typical landscape assemblage (Fig. 9.2) consists of a high **summit plain**, found at 600 m in Wales and 940 m in Scotland, surmounted by isolated mountains. Below the summit surface are two other, less well developed levels, referred to as the **middle** and **lower surfaces**. By analogy with a similar morphological situation in Brittany, where the supporting sedimentary evidence is more conclusive, it is thought that the summit surface was

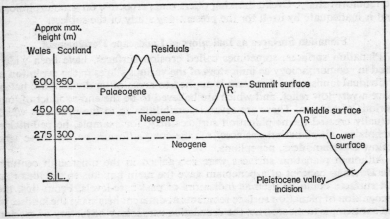

Fig. 9.2. British upland planation surfaces.

fashioned in the Palaeogene, before the Alpine orogeny; whereas the two lower surfaces postdate this mountain-building period. Pleistocene glaciation and valley incision have severely cut into all three surfaces.

These, and many of the other surfaces in Britain above about 200 m, are usually regarded as sub-aerial in origin. Benches and surfaces below this level are thought to be marine-trimmed, formed during the Pleistocene. But some of the difficulties in relying on morphological evidence alone may be illustrated with reference to the South Downs in southern England. Here, as elsewhere on the Chalk outcrop, one interpretation is that the hilltops, carrying clay-with-flints, represent a Neogene surface, and the dipslope is characterised by a staircase of Pleistocene marine benches. On the other hand, recent detailed work on the superficial deposits shows that much of the cuesta, except for the scarp face, is in fact an exhumed sub-Eocene surface passing beneath Eocene (Palaeogene) deposits in the Sussex coastal plain. No deposits of marine origin on the dipslope above the coastal plain have been found. Again, it has to be concluded that examples of denudation chronology based largely on morphological evidence are unreliable as models of landscape evolution.

Landscape Development in the Pleistocene

In contrast to the rather sketchy traces of the Cainozoic planation episodes visible in the landscape, the changes wrought by events in the latter part of the Pleistocene remain dramatic and widespread. The evidence for reconstructing landscape evolution in this epoch is sufficiently strong to allow us to determine at least a working hypothesis of events. This particularly applies to the last 50,000 years or so, which are encompassed by the range of radiocarbon dating. This technique measures the amount of carbon-14 isotope left in organic deposits, based on a constant rate of carbon decay. Organic deposits are frequently found buried beneath or lying over other deposits, such as glacial tills, and this allows non-organic deposits to be dated indirectly.

The Pleistocene in Britain and many other temperate lands was a time of active glacial and periglacial activity, major slope changes and the initiation of new drainage lines. Large-scale oscillations of sea-level, isostatic depression and uplift, and detailed modifications to our coastline also took place. In addition, the alternation of warm and cold episodes had a profound effect on our flora and fauna, and on the activities of prehistoric man. An outline of some of the biogeographical changes in the Pleistocene is considered in Chapter Twenty-Two.

The basis for the subdivision of the Pleistocene is based on climatic change (Chapter Seventeen), although the actual identification of the climatic stages rests on the evidence of fossil plants and animals, found in marine and lacustrine sediments and in peat layers. Up to the 1950s it was thought that there were only four cold stages in the Pleistocene, each of them characterised by glaciation and separated by interglacial stages. This reconstruction was largely based on the work of Penck and Bruckner in the Alps at the beginning of this century. An up-to-date summary of knowledge is presented in Table 9.1; in Britain we now recognise seven cold stages, of which the last three are known to have been glacial. The names given to these stages are based on type sites within Britain. The possible correlations with north European and Alpine events may be found in the table, but although some texts still use

Table 9.1. Cold and Temperate Stages in the Pleistocene

	Stage	Probable north European equivalents	Possible Alpine equivalent	Climate (Britain)
	FLANDRIAN			Temperate (postglacial)
UPPER PLEISTOCENE	DEVENSIAN	Weichselian	Wurm	GLACIAL
	IPSWICHIAN	Eemian		Temperate (interglacial)
	WOLSTONIAN	Saalian	Riss	GLACIAL
MIDDLE PLEISTOCENE	HOXNIAN	Holsteinian		Temperate (interglacial)
	ANGLIAN	Elsterian	Mindel	GLACIAL
	CROMERIAN			Temperate
	BEESTONIAN		?Gunz	Cold (permafrost)
	PASTONIAN			Temperate
LOWER PLEISTOCENE	BAVENTIAN		?Donau	Cold (permafrost)
	ANTIAN			Temperate
	THURNIAN			Cold
	LUDHAMIAN			Temperate
	WALTONIAN			Cold
	PLIOCENE			

Alpine names, this is not very appropriate as long-distance correlations are hazardous, and the British Isles were never at any stage affected by ice from the Alps. In fact, nowadays, the Pleistocene stratigraphy of Britain is more complete than that of the continent.

The Cold Stages

Most of the evidence for reconstruction in the preglacial section of the Pleistocene comes from the shelly crag and organic mud deposits of East Anglia. Some of these contain fossil ice-wedges and polygons, indicating that in the cold stages the climate was sufficiently severe to have induced permafrost in southern Britain. Glaciers may have formed in the northern and western highlands, although there is no direct evidence of this. Coarse gravel spreads were laid down by early courses of the Thames during this period.

In the three main glacial stages, the Anglian, Wolstonian and Devensian, each stage is known to have been made up of several ice advances, separated by slight ameliorations of climate called **interstadials**. In the **Anglian** stage, both British and Scandinavian ice-sheets contributed to the widespread deposition of till over much of East Anglia. However, any marked glacial topography has since been subdued by later erosion and this part of the country is now characterised by gently undulating till plains. Ice in the Anglian stage reached into the London Basin, diverting the Thames (see Fig. 6.10) and

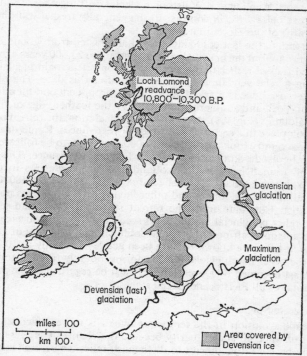

Fig. 9.3. Ice limits in the British Isles.

forming the southernmost limit of glaciation in eastern England. Elsewhere in Britain, only scattered evidence of Anglian glaciation appears to exist.

The results of **Wolstonian** glaciation are best seen in the English Midlands, where a prominent end-moraine exists at Moreton-in-Marsh. Glacial Lake Harrison was formed at this time, along with other large glacial lakes in the Bristol Channel. The mountains of Scotland, the Lake District, Ireland and Wales nourished a large ice-sheet in the Irish Sea Basin which penetrated up the Bristol Channel into the Bristol area, and reached as far south as the Isles of Scilly, pressing against the north coasts of Devon and Cornwall. The whole of Ireland was covered by ice at this time. On the other side of the British Isles, many of the striking glacial landforms of north Norfolk are thought to date from this stage. Beyond the limits of the ice-sheets, river activity created the outline of the present drainage pattern. Extensive periglacial activity deposited spreads of coombe rock in the valleys of the Chalk outcrop, and accumulations of head elsewhere.

In the present state of knowledge, both Anglian and Wolstonian ice contributed to the maximum limit of glaciation in different parts of the British Isles (Fig. 9.3). It has been suggested that ice in one of these stages reached the English channel and was responsible for moving the 'blue stones' of Stonehenge from Pembrokeshire to Wiltshire; but this view commands only limited support. Anglian and Wolstonian glacial drift together are sometimes referred to as **older drift**, in contrast to the generally fresher morphology of the **newer drift** of the Devensian stage.

The **Devensian**, the last cold stage in Britain, lasted from about 70,000 to 10,000 years ago. But during much of this time, up to 25,000 years ago, Britain experienced tundra rather than glacial conditions. The main glacial episode occurred in a relatively short space of time after this date. The maximum limit of Devensian ice is marked variously by prominent end-moraines as in the Vale of York and southern Ireland, or by the feather-edge of till sheets, as in the Midlands (Fig. 9.3). After maximum glaciation, the retreat of Devensian ice took place in a series of stillstands or readvances. Rapid deglaciation of the entire country took place about 13,000 years ago, to be followed by one final glacial episode, known as the **Loch Lomond Readvance**. This saw the creation of a small ice-sheet in Scotland and cirque glaciers in the Lake District, Wales and Ireland. It finished about 10,000 years ago (8000 B.C.).

Nearly all the eskers, kames and other fluvioglacial landforms now visible in the British Isles date from the retreat stages of Devensian glaciation. Similarly, relict periglacial features still seen at the ground surface, such as ice-wedge polygons, stripes, pingos and solifluction aprons, are of Devensian age. Some of these are known to have been active as recently as 10,000 years ago. On the other hand, glacial erosion forms, notably troughs and cirques, although last occupied by Devensian ice, must be regarded as the product of the sum total of all Pleistocene glacial stages.

The Temperate Stages

Landscape evolution in the temperate stages is more difficult to unravel than that of the cold stages, partly because the changes were less dramatic, and partly because it is difficult to distinguish relict temperate features from those created in the current temperate stage, the **Flandrian**. Botanical and zoological evidence points to climates warmer than now in some of the temperate stages, and this may account for the presence of some of our relict subtropical weathering features, such as soils of terra rossa type (see page 223) or the rounding on the tors of Devon. In the interglacial stages, the **Hoxnian** and **Ipswichian**, rivers reworked some of the freshly laid glacial and periglacial deposits in their valleys, incising into these deposits in their upper reaches and laying down new terraces in their lower courses. The major rivers of central and southern England have complex terrace sequences, a result of erosion and deposition in both warm and cold stages in the Late Pleistocene.

Further evidence of landscape evolution in the temperate stages comes from coastal areas. Raised beaches and river terraces suggest that in relative

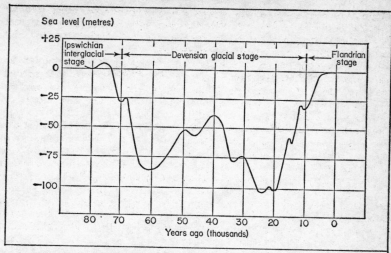

Fig. 9.4. Sea-level changes since the last (Ipswichian) interglacial.

terms, there has been a fall of the sea in relation to the land of 200 m since the beginning of the Pleistocene. In the Hoxnian interglacial, maximum sea-levels, as represented by the Goodwood raised beach in Sussex, appear to have been about 35 m higher than now. In the last interglacial, the Ipswichian, sea-levels reached 15 m, and many of the best preserved raised beaches of the Irish Sea and English Channel coasts were formed at this time at various levels up to the 15 m mark. In between the temperate interglacial stages, sea-levels dropped markedly with the growth of continental ice-sheets, reaching at least −100 m in the Devensian (Fig. 9.4). However, in the first part of our present temperate stage, the Flandrian, a major rise of sea-level took place, reaching the range of present levels about 7000 years ago. This world-wide rise of sea-level was very important in several respects. It recreated the rias on the British coastline and cut Britain off from the continent. It was also responsible for shifting a great deal of sediment landwards towards our present shore-zone, where modern coastal processes have redistributed it into sand dunes, spits and bars.

Suggested Further Reading

Davis, W. M., *Geographical Essays*, reprinted 1954, Dover Publications, New York.
King, L., *Morphology of the Earth*, Oliver & Boyd, Edinburgh, 1966.
Linton, D. L., 'Tertiary landscape evolution', in *The British Isles*, edited by J. W. Watson and J. B. Sissons, Nelson, London, 1964.
Small, R. J., 'The cycle of erosion', in *The Study of Landforms*, Cambridge University Press, London and Cambridge, 1970.
Sparks, B. W., and West, R. G., *The Ice Age in Britain*, Methuen, London, 1972.

PART TWO
WEATHER AND CLIMATE

CHAPTER TEN
THE ATMOSPHERE AND ITS ENERGY

The Nature of the Atmosphere

Most of us take the atmosphere very much for granted. We no doubt appreciate that 'weather' occurs from time to time, and that this can provide a useful topic for daily conversation, but few of us are consciously aware of the precise nature of the atmosphere, what it consists of, or of the forces that are involved in moving it about. This chapter considers some of the significant characteristics of the atmosphere and looks at the energy systems which are the basis of weather and climate.

Atmospheric Composition

The atmosphere is basically a mixture of gases held to the Earth by gravitational attraction. Although other important properties of the atmosphere, such as temperature and pressure, can vary considerably in both time and place, its composition in terms of the relative proportions of the gases present in any unit volume, tends to remain remarkably constant, at least in the lower layers of the atmosphere. Thus the atmosphere generally tends to act very much as a single gas, which we commonly know as 'air', obeying normal physical gas laws.

Table 10.1. Average Composition of Dry Air

Constituent gas	Percentage volume
Nitrogen	78·1
Oxygen	20·9
Argon	0·93
Carbon dioxide	0·03
Neon	0·0018
Helium	0·0005
Ozone	0·00006
Hydrogen	0·00005
Krypton Methane Xenon	Trace

The main component gases of dry air are listed in Table 10.1. It will be noticed that **nitrogen** and **oxygen** together make up about 99 per cent of the volume, and that the other one per cent is chiefly **argon**.

Some of the apparently minor gases have a significant role to play in the atmosphere. **Carbon dioxide** is important because of its ability to absorb heat.

This allows the layers of the atmosphere to be warmed by the sun's heat and by radiation coming from the surface of the Earth. Because of this property, it has been suggested that varying amounts of carbon dioxide in the atmosphere in the past could be one of the causes of climatic change (Chapter Seventeen). **Ozone** is another gas that acts similarly in absorbing radiation.

It must be emphasised that Table 10.1 refers to the average constituents of *dry* air. The lower parts of the atmosphere, up to 10–15 km, contain in addition **water vapour**, which is largely derived by evaporation from water bodies on the earth and by transpiration from plants. It is not found at great heights in the atmosphere, partly because mixing and turbulence is not sufficiently strong to carry it up very far, and partly because the upper atmosphere is too cold to absorb it. Water vapour, too, is capable of absorbing heat, but even more important than this is the fact that its presence in the atmosphere is fundamental to many essential meterological processes, such as rain-making. These will be explained in the next chapter.

The atmosphere also carries in suspension variable amounts of **solid material**, which takes the form of minute dust particles derived from natural agencies or from man-created pollution. Tiny salt particles introduced by evaporation over oceans are also present. This fine material provides the necessary nuclei on which water vapour can condense to form water droplets and eventually precipitation. Large amounts of dust tend to make the atmosphere hazy, and in extreme cases, where pollution is involved, particles in the atmosphere can be positively harmful to health.

Variations in Composition with Height

Constant mixing and turbulence in the lower regions of the atmosphere help to prevent any of the constituent gases, especially the lighter ones, from separating out to form individual layers. However, at great heights marked concentrations of certain gases do occur. Ozone is found concentrated at levels between 15 and 35 km above the Earth's surface. The process that creates ozone, the separation and recombination of oxygen molecules in different form, is most marked here. Although the amounts involved only represent a slight increase on the very small amounts of ozone normally found elsewhere, the concentration does lead to an increase in atmospheric temperature at this particular level.

At even greater heights, beyond about 100 km, where the atmosphere is very thin and turbulence and mixing apparently non-existent, recent data from satellites and rockets suggest that the lightest gases do in fact separate out, forming several concentric gas envelopes around the Earth. The innermost of these is a nitrogen layer, found at heights between 100 and 200 km; this is succeeded in turn by layers of oxygen (200–1100 km) and helium (1100–3500 km); and finally beyond 3500 km only hydrogen exists, to which there is really no clearly defined upper limit.

Atmospheric Mass

It is not always realised that the atmosphere has mass and weight, to the extent of several millions of tons. At sea-level, this weight expresses itself by exerting a pressure of about 1 kilogram per square centimetre (15 lb./sq. in.) on any surface. We are not normally aware of this pressure because it acts in all directions; for instance, in an empty container, the pressure on the outside is counterbalanced by the pressure of air on the inside, otherwise the container would collapse.

One way of measuring pressure is with a mercury barometer. Torricelli's famous experiment of 1643 showed that air pressure is able to support a column of mercury 762 mm high in a glass tube (Fig. 10.1). The modern metric unit of pressure measurement is the millibar (mb), one millibar being equal to the pressure necessary to support 0·75 mm of the mercury column.

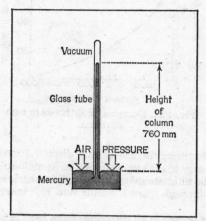

Fig. 10.1. Torricelli's experiment to illustrate the effect of air pressure.

Air is an extremely compressible gas, and as a result atmospheric pressure varies considerably with height. Since the pressure will vary according to the weight of air above it, the lowest layers of air are densest. The mean sea-level pressure of the atmosphere is 1013·2 mb; at 6 km (17,000 ft) pressure has decreased by half, and this is the approximate limit of normal human habitation on Earth. The rate of pressure decrease is not constant, however (Fig. 10.2). Near sea-level it is approximately one millibar every 10 metres, but this rate gradually lessens, and in the tenuous upper atmosphere it is very slight indeed.

Atmospheric Energy

Energy from the Sun

So far, we have been considering the atmosphere largely as if it were static, but in reality, as we know from the constant passage and change of weather elements, it is very much a dynamic entity. Large volumes of air are continually being moved both up and down and across the face of the Earth. Clearly a great deal of energy is involved here, and in order to understand weather systems, it is necessary that we know something about the sources of all this power.

It should be appreciated that the atmosphere is not a closed energy system. It is in contact with both the Earth and with space, and receives energy from

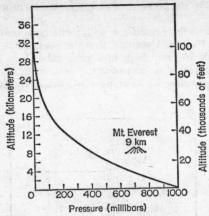

Fig. 10.2. Decrease of air pressure with
elevation.

both directions. However, the Earth itself directly contributes only a negligible amount of energy to the atmosphere, and its main role is to reflect energy from elsewhere. The ultimate sole source of atmospheric energy is in fact heat and light received through space from the sun. This energy is known as **solar insolation.**

The sun radiates energy in all directions from itself, and the Earth receives only a tiny fraction of the total output. The unit of measurement of this energy is the **langley,** one langley being equal to one gram-calorie per square centimetre. Measurements by satellites at the top of the atmosphere indicate that the radiation rate from the sun is two langleys per minute. This amount of energy is termed the **solar constant,** and it appears to vary little in intensity from year to year. It is perhaps marginally affected by sunspot cycles and variations in the Earth's orbit.

Insolation over the Globe

Substantial variations in the amount of insolation received over the face of the globe occur because the Earth is a sphere. If we assume for the moment that the Earth has no atmosphere, only where the sun is directly overhead will the energy reception rate on the Earth's surface be two langleys per minute. Elsewhere, because of the Earth's curvature, the same amount of insolation will be spread over much greater areas. Hence **latitude** is an important factor in determining the amount of energy received, and on this basis the polar regions will generally receive far less insolation than equatorial areas.

Length of day, which is itself partly controlled by latitude and partly by the season of the year, is a second factor influencing the amount of energy available, since clearly the longer the sun shines the more the insolation received. Thus in the summer months, the lack of intensity of radiation received by the poles is to some extent counterbalanced by its long continuity. However, taking into account the long polar night, the total amount of insolation received by polar regions is still considerably less than in lower latitudes. A perspective diagram (Fig. 10.3) shows the theoretical receipt of insolation over the globe for various times of the year.

To these considerations we must add a third factor, that of the effect of the atmosphere itself. Important changes take place to insolation once it enters the atmosphere, and these we can now examine.

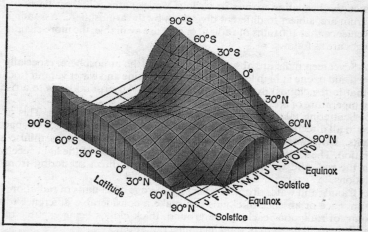

Fig. 10.3. Variations of insolation with latitude and season. The amount of energy received at any point is proportional to the height of the point above the base of the block.

The Atmospheric Energy Budget

We have seen that the atmosphere is in constant receipt of solar heat and energy. Yet by and large the atmosphere of the Earth is not getting any hotter. We have to conclude from this that there must be some sort of energy balance in the atmosphere whereby the amount of energy coming in is balanced by equal losses, otherwise temperatures would simply rapidly rise to become intolerable.

The balance is achieved by a complex series of energy transfers, and at least three common types of energy transport are involved in these transfers. The most important is **radiation**, which we have already encountered as the means by which insolation reaches us. Radiation is the transference of heat by electromagnetic waves. The electromagnetic spectrum includes X-rays, heat rays, light rays and radio waves. The wavelength at which a body radiates depends on its surface temperature. Thus the sun, having an extremely hot surface temperature (about 5000°C), radiates fairly short wavelengths, part of which are felt as warmth, part of which are visible as light. The Earth, on the other hand, having a cool surface, re-radiates heat at much longer wavelengths.

Heat is also transferred in the atmosphere by **convection**, which involves the mass movement of gas or liquids, the heat acquired by the liquid or gas being transported with the medium. The third main method is by **conduction**, which is the transference of heat by actual contact. For example, if one end of a poker is left in the fire, the other end will soon become hot by conduction. Since air is rather a poor conductor, the transfer of heat by conduction is most important in the lowermost layers of air in contact with the ground, and in the ground itself.

Atmospheric Response to Incoming Radiation

Since the insolation arrives in various wavelengths, different parts of the spectrum are subject to different diversions by the atmosphere. Assuming for convenience that 100 units of radiant energy are available, the more significant changes are as follows.

(*a*) Seventeen units are absorbed by gases in the atmosphere, especially by oxygen and ozone at high levels, and carbon dioxide and water vapour (including that in the clouds) in the lower atmosphere. Absorption leads to a rise in the temperature of air.

(*b*) Scattering takes place by gas molecules and dust particles. This takes place in all directions, some of it Earthwards, but the net loss back to space is 6 units. Downwards-directed scattering is generally known as diffuse sky radiation. The scattering is more effective at the blue end of the light spectrum than the red, and we see the sky as blue from indirect scattering from all directions.

(*c*) Clouds and water droplets reflect an average of 23 units of radiation, but the presence or absence of clouds can make a considerable difference to the amount of radiation reaching the ground, thick clouds being capable of reflecting up to 80 per cent of total incoming radiation.

(*d*) Reflection of radiation also takes place from the Earth's surface. Generally this amounts to 7 units, but it varies with the nature of the ground. Water, which makes up much of the Earth's surface, has only a low reflection value or **albedo**, whereas land surfaces have a much higher albedo.

All the changes mentioned so far are accomplished by radiation. The total amount of energy lost by scattering and reflection of various kinds and returned to space (i.e. $b+c+d$) is usually termed the **Earth's albedo**, and the amount involved is 36 units. If we add to this the 17 units directly absorbed by the atmosphere, it will be apparent that less than half (47 units) of the original insolation received at the top of the atmosphere actually gets through to the ground.

Terrestrial Radiation

The energy received by the Earth is re-radiated at much longer wavelengths back into the atmosphere. Some of this terrestrial radiation is lost directly into space (8 units). However, the atmospheric gases are able to cope much more easily with long-wave radiation than short, and a great deal of the rest of the radiation is absorbed by the atmosphere, especially the clouds. The atmosphere itself again re-radiates or reflects much of this heat back again, setting up a continuous interchange of energy with the ground. In this process there is a net gain to the atmosphere of about 14 units.

Further heat is lost from the Earth to the atmosphere in the form of latent (stored) heat when evaporation occurs. The heat is carried upwards in convection currents and released to the atmosphere on condensation (see Chapter Eleven for further details of the this process); 23 units are involved here.

There is also a small amount of conduction of heat between the ground and the atmosphere. Any heat transferred this way is again carried upwards by turbulence or convection. Although the atmosphere itself can warm the ground by conduction, it is thought that there is a net gain by the atmosphere of two units.

The heat gained by the atmosphere from the ground amounts to 39 units; this joins the energy units gained by the atmosphere from incoming solar radiation in being eventually radiated by the atmosphere back into space (56 units in all). If we add to this the energy losses associated with the Earth's albedo (39), plus the direct radiation loss from Earth (8), we achieve 100 units again to complete the balance. This system is represented diagramatically in Fig. 10.4.

One of the most significant facts in the energy budget is that the atmosphere is largely heated from below. The atmosphere either reflects or lets through most of the incoming short-wave radiation, absorbing only a small part of it. On the other hand, it traps a great deal of the outgoing terrestrial energy, and by this means the atmosphere is warmed. Much the same principle is employed in a greenhouse; the glass lets in insolation but does not allow the warm air inside to escape so readily. Hence the term **'Greenhouse Effect'** is often used to illustrate one of the fundamental heating principles in the atmosphere.

Horizontal Energy Transfer

We have been dealing with a budget situation in a vertical sense: up and down movements of energy. The figures quoted represent the average for the whole of the Earth's atmosphere. It can also be demonstrated that there is an important second type of budget in the atmosphere in which *horizontal* movements take place. This is because the vertical budget varies considerably over the globe, and horizontal movements take place to compensate these differences.

Earlier we saw that there was some imbalance in the amount of energy received over the globe. The effect of the atmosphere is to emphasise this imbalance. At the equator, where the sun is overhead, the incoming radiation has only the vertical thickness of the atmosphere to penetrate, and a fairly large proportion gets through. At high latitudes, since the radiation approaches obliquely, it has more atmosphere to penetrate and is more liable to scattering and reflection; polar regions also tend to be more cloudy. On the other side of the budget, less outgoing radiation is in fact lost by the poles than the equator, but this is not enough to make up for the incoming energy deficit.

The net result is demonstrated in Fig. 10.5: the equatorial areas have a heat surplus (positive budget), whereas the poles have a negative budget. This in theory should mean that tropical areas should get steadily warmer, and the Arctic and Antarctic even colder. But such is not the case, as the mean temperatures of both areas remain fairly constant. This is explained by the presence in the atmosphere of large horizontal circulation systems, whereby the excess heat received at low latitudes is transferred to the poles. The energy is transported through various media, including ocean currents and wind systems. This transference of energy from equator to poles is one of the fundamental driving forces behind the general circulation of the atmosphere around the globe (Chapter Thirteen).

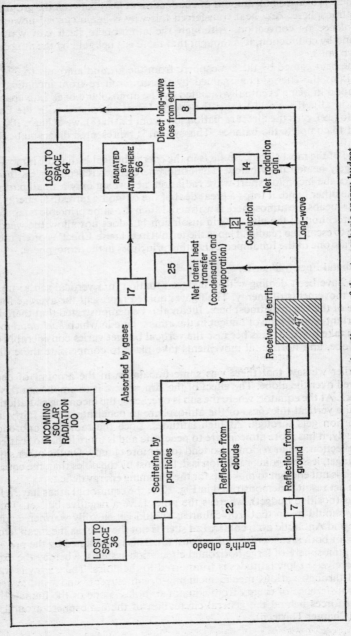

Fig. 10.4. A flowline relation diagram of the atmospheric energy budget.

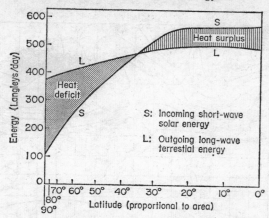

Fig. 10.5. Balance between incoming and outgoing
radiation for various latitudes.

(After A. N. Strahler, courtesy
Harper and Row)

Air Temperatures

We may now briefly illustrate some of the broad effects of atmospheric
energy on air temperature patterns. Air temperatures depend on a large number of geographical factors, including elevation, aspect, proximity to the sea,
and direction of prevailing wind. Even more fundamental than these is the
strong correlation they show with patterns of insolation.

Annual Temperature Patterns

On maps of global air temperatures for any time of year (Figs. 10.6 and
10.7) one can usually distinguish east–west temperature zones in which the
isotherms (lines of equal temperature) run more or less parallel to lines of
latitude. This pattern basically reflects the general decrease of insolation from
equator to poles. During the year, the isotherms mirror changing air temperatures by moving northwards and southwards following the declination of the
sun.

The pattern of temperature change is not the same everywhere, as representative graphs in Fig. 10.8 show. In equatorial areas, where the diurnal
receipt of insolation is fairly constant all year round, the annual temperature
cycle shows little seasonal variation. In mid- and high-latitude regions it is
much more marked: in areas between the tropic and polar circle in each
hemisphere the sun's path shifts through a relatively large range of noon altitudes, and substantial seasonal differences exist in the length of day. Thus
insolation amounts and, correspondingly, air temperatures show a marked
seasonal pattern. In polar regions there are very large seasonal contrasts in air
temperature, reflecting the difference between the large negative energy budget
of the long polar night on the one hand, and the continuous period of summer
insolation on the other.

Despite the strong general correlation between the insolation pattern and

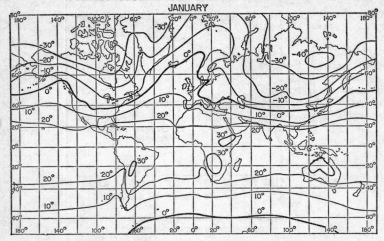

Fig. 10.6. Mean January temperatures (°C).

the air temperature pattern, there is frequently a marked timelag between the two, sometimes to be measured in terms of several weeks. This serves to remind us again that the atmosphere is not directly heated by insolation, but only indirectly by it through the medium of the ground. In the northern hemisphere, it is July, and not June, which is the warmest month. The lag occurs because the ground continues to warm up well past the actual peak of insolation at the summer solstice. Similarly, in winter the coldest air temperatures are usually experienced in January, some time after the winter solstice,

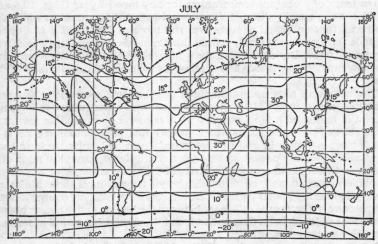

Fig. 10.7. Mean July temperatures (°C).

as the ground continues to lose heat even after insolation has begun to increase.

The lag in temperature is often much greater over oceans than over continents. This is because land and sea have different heating qualities which can have a marked effect on the energy budget. Land is much more responsive to heat; it warms and cools quickly, and experiences more extreme temperatures. Land has a low **specific heat**, i.e. it requires relatively few heating units to make it warm up by a unit temperature. Sea areas, on the other hand, have a high specific heat; they require a considerable amount of heat to make them change by a unit temperature, and they respond to changes much more slowly. They also store heat by transporting it downwards in currents to mix with deeper waters.

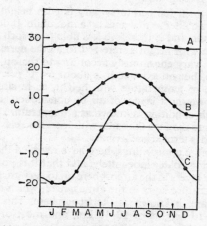

Fig. 10.8. Contrasting annual patterns of air temperature (A Colombo, Sri Lanka; B Kew, England; C Hebron, Labrador).

The general implications of these differences is that climatic stations in the middle of continents tend to have fairly marked seasonal differences of temperature which follow insolation patterns with only a short timelag. Maritime stations have much more equable seasonal temperatures and frequently their warmest and coldest periods may be up to two months after the solstices.

Diurnal Changes of Temperature

The daily pattern of temperature change that we normally experience illustrates energy changes on a small time-scale. Since the atmosphere is largely heated from the ground, as long as the ground is the colder of the two, the air temperature will fall. On a calm day with little cloud, air temperatures usually reach their minimum just before sunrise, because the ground has been giving off long-wave radiation all through the night, gradually becoming colder and cooling the air above by conduction. Within about half an hour of sunrise, incoming solar radiation causes the temperature of the ground to rise, and in response to this the air temperature also begins to rise with a lag of about an hour. A positive budget situation develops in which the incoming energy exceeds the outgoing and therefore temperatures must inevitably rise.

Maximum insolation is received at midday, but air temperatures still continue to rise beyond this time, because the ground is still gaining heat from the early afternoon insolation. The peak of air temperature is usually about 14.00 hours. After this time, although there is still a radiation surplus, increased turbulence associated with convection currents mixes cooler upper air with warm air near the ground, causing temperatures to drop. After sunset, the air initially remains fairly warm as it is still being heated by long-wave radiation from the ground, but this gradually expires and temperatures drop accordingly.

Vertical Changes of Temperature

Another way in which temperatures vary is with height. It is commonly observed that air gets colder as one goes up a mountain. It does so because the air becomes less dense, and is therefore less able to absorb heat. The normal drop of temperature with height is referred to as the **normal lapse rate**. The lapse rate of air can vary enormously according to season, time of day, and geographic position, but on average it is about 6·4°C per kilometre. Where there is an increase of temperature with height, as in situations where the ground cools the lowermost layers of air by conduction, leaving the higher layers still warm, this condition is known as a temperature **inversion**. In cases where the temperature remains the same with increased elevation, the layer of atmosphere involved is termed **isothermal**.

Explorations into the upper atmosphere have revealed that the atmosphere does not go on getting colder indefinitely, and that it is possible to recognise distinct temperature layers in the atmosphere up to very great heights.

The lowest temperature layer in the atmosphere, which is characterised by the normal lapse rate, is known as the **troposphere**. This extends from the Earth's surface up to heights of 10–15 km, and most of the weather that materially affects us takes place within this zone.

At the top of the troposphere, there is a reversal in the temperature gradient known as the **tropopause**. This inversion acts as a ceiling to weather generated in the troposphere, which in many respects is therefore self-contained. The actual height of the tropopause varies latitudinally and seasonally. It tends to be higher in elevation in summer than in winter, and higher at the equator than at the poles.

The next major temperature layer is the **stratosphere**. This is generally regarded as having two parts. The lower part immediately above the tropopause is largely isothermal, or increases only slowly in temperature with height, but in the upper or 'warm' stratosphere, beyond heights of 30 km, there is a more rapid rise of temperature. The main reason for this is the effect of the concentration of ozone at this level.

Layers above the stratosphere are less easy to define in terms of temperature. In Fig. 10.9 the **mesosphere** (50–80 km) is shown as a region of falling temperatures, separated from the **thermosphere** (80–500 km) by another inversion, the mesopause. Temperatures, theoretically, rise in the thermosphere, but the atmosphere is so tenuous here that sensible air temperatures mean very little.

The thermosphere is coincident with the **ionosphere**, for it is here that indivi-

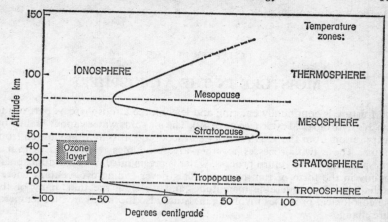

Fig. 10.9. Layers in the atmosphere.

dual molecules of gas become electrically charged or ionised. The best known of the several ionic layers is the Heaviside layer, which reflects radio waves.

The term **exosphere** is sometimes applied to the outermost part of the Earth's gas atmosphere which exists beyond the thermosphere at heights of 500–2000 km. Beyond this, we are in the realm of the **magnetosphere**, where the influence of the Earth's magnetic field is still felt, but there is no atmosphere.

Suggested Further Reading

Barry, R. G., and Chorley, R. J., *Atmosphere, Weather, and Climate* (Chapter 1), Methuen, London, 1970.

Dobson, G. B. M., *Exploring the Atmosphere* (3rd edn.), Oxford University Press, Oxford and London, 1972.

Lamb, H. H., *Climate: Past, Present and Future* (Vol. 1, Chapter 2), Methuen, London, 1972.

Parker, E. N., 'The solar wind', *Scientific American*, Vol. 210, pp. 66–76, 1964.

Peterson, J., *Introduction to Meteorology* (Chapters 2 and 3, 3rd edn.), McGraw-Hill, New York, 1969.

CHAPTER ELEVEN

MOISTURE IN THE ATMOSPHERE

Moisture is continually entering and leaving the atmosphere as part of the Earth's hydrological cycle (see page 42). The air gains water vapour mainly by evaporation from the Earth's water surfaces, and also by **transpiration** from plants. These two processes together are sometimes referred to as **evapotranspiration**. In the return process, water is **precipitated** from the atmosphere mainly in the form of rain or snow. This never-ending cycle entails energy exchanges of considerable magnitude, and we shall see that many of the meteorological processes involving moisture also inevitably involve temperature changes.

Variations in atmospheric moisture content depend on a number of factors, but probably the most important control is that of temperature. This is because temperature fundamentally affects the moisture-holding capacity, or **saturation limit**, of the air. Cold air can hold very little moisture, and consequently the water vapour content is always low in absolute terms when the air temperature is low. As the temperature is increased, so the moisture-holding capacity also increases; in other words, much more moisture can be evaporated into warm air than cold air before it becomes saturated. For instance, air at 20°C can hold four times as much water by weight as air at freezing point.

Humidity and its Expression

There are several different ways of describing the amount of moisture in the air. One way is in terms of **water vapour pressure**, stated in millibars. Water vapour, like any other gas, exerts a pressure which contributes to the sum total of atmospheric pressure. Following from this is the **saturation vapour pressure**, which is the pressure exerted by water vapour when the air is saturated. This will vary with temperature, and the nature of the relationship is illustrated in Fig. 11.1 in the form of a graph. It will be noticed that for air temperatures below 0°C, air which is saturated in respect to water (supercooled) is more than saturated with respect to ice. We will recall this point later when dealing with precipitation processes.

The **absolute humidity** of the air is the mass or weight of vapour per unit volume of air, usually calculated in grams per cubic metre. This is a direct way of expressing humidity, but perhaps more useful is the term **specific humidity**, which is the ratio of the weight of water vapour (grams) to the weight of the moist air (kilograms). An average specific humidity for tropospheric air in mid-latitudes, for instance, would be about 10 g/kg. One advantage this term has over absolute humidity is that when air is lifted to higher elevations without gain or loss of moisture, the specific humidity, being concerned with weight, remains constant, whereas the absolute humidity is affected by the volume change and alters progressively. An alternative sometimes used to specific humidity is the **mixing ratio**, which is defined as the weight of water vapour to the weight of dry air (not including the water vapour). Numerically this expression is almost identical to the specific humidity.

The terms just described are in common use in meteorology, but probably more familiar to the layman is the expression **relative humidity**. It defines the ratio of the actual amount of water vapour in the air to the maximum amount the air could hold at that temperature, expressed as a percentage, i.e.

$$\text{Relative Humidity (R.H.)} = \frac{\text{Actual amount}}{\text{Possible amount}} \times 100\%$$

Thus air which is completely saturated has a relative humidity of 100 per cent; if it contains only half the amount of water needed to saturate it, then the relative humidity is 50 per cent.

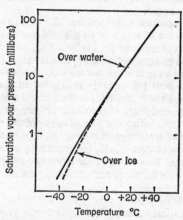

Fig. 11.1. Relation between saturation vapour pressure and temperature.

There are two basic ways of changing the relative humidity of an air mass. One is to change the water content: if the air receives more water by evaporation, then, providing the temperature is kept constant, the relative humidity will increase. The other way is to alter the temperature: with the moisture content constant, if the temperature of an air parcel is raised, then the relative humidity will fall. This is because the potential capacity of air to hold water has increased, thus altering the R.H. ratio. Conversely, if the temperature is decreased, the relative humidity automatically rises. Thus on a diurnal basis, even with no actual moisture changes, the relative humidity can often change quite markedly in response to the daily pattern of temperature. On a fine day it may be quite low around midday, perhaps 40 per cent, but at night falling temperatures may cause it to rise to 90 per cent, or even to saturation point, in which case condensation may occur in the form of dew or fog. Related to this process is the term **dew-point** temperature, which can be used to denote the temperature at which condensation would occur if the air was cooled sufficiently.

Relative humidity is the most useful everyday way of referring to humidity;

among other things, human comfort depends on it. However, the expression does have its disadvantages. Suppose, for example, we have an air mass with a temperature of 20°C and a relative humidity of 'only' 60 per cent. It is in fact carrying far more moisture than one with a temperature of 5°C and a relative humidity of 95 per cent. In other words we cannot use relative humidity for direct comparison between air masses.

Evaporation and Condensation

So far we have been largely concerned with a consideration of water in its vapour state. However, moisture exists in all three states of matter in the atmosphere: as water vapour, in liquid state as water, and in solid state as ice. Changes from one to the other are frequent occurrences in the atmosphere and are known as **phase changes** (Fig. 11.2). A certain amount of energy is used in accomplishing phase changes.

In the process of **evaporation**, something like 600 calories of heat are required to change one gram of water from a liquid to a vapour state. The effect of this on one kilogram of air is to cool it by 2·5°C. Normally such a heat loss is quickly compensated by conduction and radiation.

The heat loss during evaporation passes into the water vapour in hidden form known as **latent heat** (of vaporisation). When the reverse process of **condensation** takes place, the latent heat locked in the water vapour is released back to the atmosphere, causing a slight rise in air temperature.

Similar heat exchanges, although not always to the same degree, are involved in the processes of freezing and melting, and sublimation and deposition (Fig. 11.2). **Sublimation** occurs when, for example, snow patches disappear without first melting, and **deposition** of ice from vapour in the form of rime is quite common in very cold climates and at high altitudes.

Factors Favouring Evaporation

Technically, evaporation occurs when the vapour pressure at a water surface exceeds that in the atmosphere above. Evaporation tries to equalise the

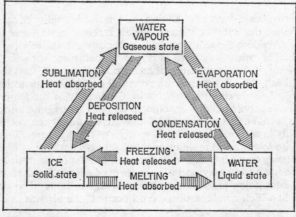

Fig. 11.2. Water phase changes and latent heat exchanges.

pressure towards saturation levels so that it is the same in the air as at the water surface. This is the basis of three principal factors which favour evaporation.

First, evaporation depends on the **initial humidity** of the air. If the air is very dry, and consequently the atmospheric vapour pressure low, then strong evaporation is likely to occur. On the other hand, little evaporation will occur if the air is almost saturated. As a general rule, the drier the air, the greater the evaporation from water surfaces. A second requirement is that a **supply of heat** is necessary to maintain evaporation. The rate of evaporation will be proportionally higher depending on the warmth of the water surface and the air immediately above it.

Third, **wind strength** can have a considerable effect on evaporation. In perfectly calm conditions, evaporation quickly saturates the overlying air, thus limiting the process. But in turbulent conditions, the saturated air is being continually replaced by fresh air. Washing on a clothes-line therefore dries much more quickly on a windy day than on a calm one, and in general we can say that the greater the wind strength, the more effective evaporation is likely to be.

Factors Favouring Condensation

We have seen that condensation is liable to occur either when enough water is evaporated into the air mass for it to reach saturation, or alternatively when the temperature drops sufficiently for it to reach dew point. The first method is in fact quite rare, and much more common are circumstances favourable to condensation by cooling. These include the radiation cooling of the air; contact cooling of the air when it rests over a cold surface (usually at night); the mixing of warm, moist air masses with cooler air; and adiabatic or expansion cooling (see below).

There is one further factor to consider. For some time now it has been known that condensation occurs only with the utmost difficulty in pure air. In the higher layers of the atmosphere, where the air is likely to approach greater purity, it is possible to have relative humidities of well over 100 per cent without condensation occurring. This is because there must be some tiny particle or nucleus of a non-aqueous substance on which the water vapour can condense in order for the process to operate. In the lower atmosphere, this normally presents little problem, since suitable condensation nuclei exist in great profusion. The main substances which act in this way include common salt derived from the sea, dust, and pollution particles. Some of these nuclei are very **hygroscopic**, that is they are 'water-seeking', and condensation may be initiated on them when the relative humidity is as low as 80 per cent.

Adiabatic Processes

The process of adiabatic or expansion cooling depends in the first instance on parcels of air rising through the atmosphere to higher elevations. This is a frequent occurrence in the atmosphere and the consequent cooling of the air is responsible for the initiation of much of the condensation occurring at all levels. The main ways in which air may be induced to rise are: by **convection**, caused by heating of the ground below; by **orographic** uplift, where air is forced to rise over hills and mountains; by **turbulence** in the airflow; and by uplift at **frontal surfaces** (Chapter Fourteen). Under any of these circumstances when air moves from one level to the other, temperature changes automatically ensue. The decrease of pressure with height allows the rising air parcel to expand. In accomplishing the expansion, energy is used up, and this has to be provided for within the air parcel. It is worth recalling here that air is a poor conductor of heat; thus only a negligible amount of heat will be transferred from surrounding air. In compensation for the work done in the expansion process, when a mass of air is moving to a lower level, it gains heat by contraction. These temperature changes, involving no external heat exchange, are termed **adiabatic**. They can be distinguished from non-adiabatic (**diabatic**) temperature changes which involve the physical mixing of air.

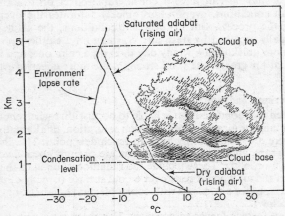

Fig. 11.3. Unstable air conditions, with air rising first at dry and then at saturated adiabatic lapse rates. Ascent ceases when the rising air has the same temperature as its surroundings.

Adiabatic cooling and warming in dry air takes place at a fixed rate of 10°C/km. This is known as the **dry adiabatic lapse rate**. For air in which condensation is occurring, the **saturated adiabatic lapse rate** pertains. This has lower values, between 4°C/km and 9°C/km, because latent heat released in the condensation process partly offsets the adiabatic temperature loss. The rate varies because the amount of latent heat released will be much greater for warm saturated air than for cold saturated air, which has a lapse rate more nearly approaching that for dry air.

It is important not to confuse the two adiabatic lapse rates with the normal or **environmental lapse rate** of the atmosphere discussed in the previous chapter. The latter is the actual temperature decrease with height as might be recorded by an observer ascending in a balloon. The adiabatic rates are dynamic, i.e. they are followed only when air is actually in the process of moving up or down.

These various lapse rates can be represented on a simple temperature/height diagram as in Fig. 11.3. Here we can follow the behaviour of a small body of rising air which has a temperature near the ground of 20°C and a dew-point temperature of 10°C. When the air is first lifted, it cools dry-adiabatically. On reaching its dew-point temperature, it cools from then onwards at the saturated adiabatic rate. The saturation level is also approximately the condensation level, i.e. the base of the cloud, in this case at about 1 km. The air continues to rise at the saturated rate until such time as it reaches the same temperature, and therefore the same density, as its surroundings, whose temperature is represented by the environmental lapse rate curve. This level will mark the limit of cloud development.

One very notable example of the operation of adiabatic processes is provided by the **Föhn** (European Alps) and **Chinook** (Alberta) winds (Fig. 11.4). These winds approach the mountains as fairly warm moist air streams, which, on being forced to rise, quickly reach condensation level and therefore lose temperature for the greater part of the ascent at the saturated adiabatic rate. By the time the summit is reached, most of the moisture has been lost, and

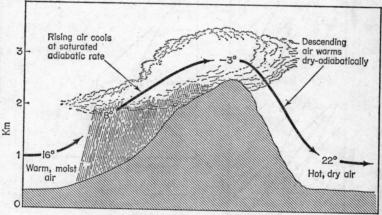

Fig. 11.4. The Föhn (Chinook) effect.

thus on descending the lee side of the mountain, these winds warm up rapidly at the dry adiabatic rate. They reach the plains or valley bottoms as hot winds with low relative humidities, and can have a drastic effect by clearing away snow very rapidly.

Stability and Instability

In discussing adiabatic changes, it has been assumed so far that the air continues to rise, even though the original cause of the uplift may have ceased to be effective. Often the air sinks back to its former level; only in certain cases will the air continue to rise by itself. This introduces the concepts of **air stability** and **instability**, which describe the buoyancy characteristics of air.

Air is defined as **unstable** if the environmental lapse rate exceeds the dry adiabatic lapse rate. This is so in Fig. 11.4, where near the ground the atmosphere has a lapse rate of about 12°C/km. In this sort of situation, if a small body of air is displaced slightly upwards by some means, it rises dry-adiabatically and immediately becomes warmer and lighter than its surroundings. It therefore continues to rise spontaneously. Absolutely unstable air conditions tend to occur on very hot days, when the ground layers of air are considerably heated, giving a higher lapse rate. If the air is moist enough, strong vertical cloud development is likely.

Stable air conditions (absolute stability), on the other hand, exist when the environmental air has a lapse rate that is less than both the dry and saturated adiabatic rates (Fig. 11.5a). In this situation, if a parcel of air is displaced upwards, whether it is dry or saturated, it immediately becomes cooler and denser than its surroundings and will sink to the ground again. The only

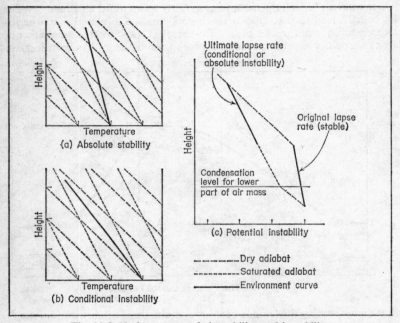

Fig. 11.5. Various states of air stability and instability.

circumstances where stable air can be made to rise is by forced ascent—for example, over high ground.

There are two special conditions of stability. **Conditional stability** (Fig. 11.5b) exists when the environmental lapse rate lies between the dry and saturated adiabatic values. This air is stable in respect to the dry rate, and would normally sink to its original level if displaced. However, if the air should become saturated, perhaps by being forced to rise to higher elevations, then on rising at the saturated adiabatic rate it would become warmer than the environment air and would continue to lift of its own accord. Thus the air is unstable, conditional on it being saturated.

Potential instability (Fig. 11.5c) is a condition which exists when large air masses which are moist in their lower layers but dry in their upper, undergo bodily lifting. This may happen, for example, at a frontal surface or on approaching higher ground. The air may initially be stable, but on lifting, whereas the dry upper parts of the air mass will rise dry-adiabatically, the lower moister parts quickly become fully saturated on rising, and cool at the saturated adiabatic rate. Thus, with different parts of the air cooling at different rates, this may radically change the temperature distribution throughout the mass, transforming an initially stable situation into an unstable one. The situation of potential instability is fairly common among warm air masses which have picked up a good deal of moisture in their lower layers on passing over sea areas.

Clouds and Fog

Clouds consist of a great number of tiny water droplets, and are a visible manifestation of condensation on a fairly large scale. Everyone is aware that clouds occur in many shapes and sizes, and vary from white to black, depending on how thick they are and whether the sun is shining on them or not. Because of the profusion of cloud forms, some sort of cloud classification is obviously necessary if we are to make some rational sense of them. One can usually draw a distinction between **stratiform** clouds, which have a layer-like appearance, and **cumuliform** clouds, which are heaped or massive in shape. To some extent this is incorporated in the internationally accepted classification which is contained in the International Cloud Atlas of the World Meteorological Organisation. In this, ten cloud 'genera' are described, which can be grouped into four families:

High Clouds	Cirrus Cirrostratus Cirrocumulus
Middle Clouds	Altostratus Altocumulus
Low Clouds	Stratus Nimbostratus Stratocumulus
Clouds with vertical development	Cumulus Cumulonimbus

Each of these main types can be divided into various 'species' and varieties, giving a very wide range of detailed forms. Fig. 11.6 gives an impression of the shapes of the principal clouds.

The clouds in the highest group are composed largely of ice crystals. This group includes the wispy **cirrus**, and the popularly-designated mackerel sky, **cirrocumulus**. **Cirrostratus** produces a halo effect around the sun or moon. The prefix **alto-** defines the middle clouds, and these are generally found at heights between 3 and 6 km. These clouds are usually formed of water droplets, which frequently exist in a supercooled state at temperatures well below freezing. Perhaps the most obvious feature of the third family of clouds, the low clouds, is that they are often indicative of dull weather. **Stratus** is a dense grey low-lying cloud, and **nimbostratus** is of a similar type, except that the prefix **nimbo-** indicates the presence of precipitation. **Stratocumulus** commonly represents

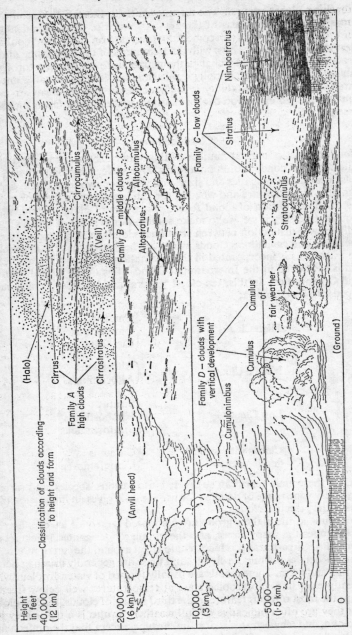

Fig. 11.6. Cloud types, grouped into families according to height, range and form.

the break-up of stratus cloud, taking the form of long cloud rolls, with clearer sky in between.

The vertically developed clouds reflect in their shape something of the up-currents within them. **Cumulus** is the familiar white woolpack cloud, typically flat-bottomed and with billowing upper parts. This cloud is often indicative of bright brisk weather. **Cumulonimbus**, which can develop out of cumulus, is on the other hand associated with heavy precipitation and thunderstorms.

This largely descriptive classification of clouds is objective and of universal application. However, it does not tell us very much about the conditions under which the clouds were formed. Two related factors are important in actually determining cloud shape, that of air stability and the mode of uplift.

In unstable conditions, the dominant form of uplift is frequently convection, and this is primarily responsible for the vertically developed clouds. Stratiform types, on the other hand, tend to be the product of stable air conditions, in which turbulence is the principal cloud-forming mechanism. Frontal uplift gives rise to a variety of clouds, depending on the type of front and the stability of the air mass involved (Chapter Fourteen). Similarly, clouds formed by orographic uplift can be either stratiform or cumuliform, depending on the stability of the air. Two interesting special types of cloud formed by orographic uplift under stable conditions are the **banner** and **wave clouds** (Fig. 11.7). Banner clouds are formed when an uplifted moist airstream reaches condensation level only at the very summit, where a small cloud forms. Further downwind in the lee of the hill the air sinks again and the cloud dissipates. Wave clouds likewise reflect the influence of topography on the flow of air.

The Formation of Fog

Fog is simply cloud that forms close to the ground. The names given to various types of fog are indicative of the way in which cooling took place to give rise to condensation.

Radiation fog is associated with radiation cooling of the land at night. In turn, the ground chills the adjacent air layers by conduction. This type of fog

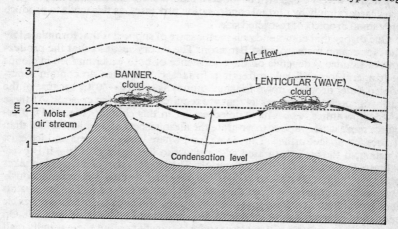

Fig. 11.7. Banner and lee clouds.

is most likely to form under fairly calm conditions with clear skies, which allow maximum outgoing radiation. Autumn and early winter are usually the seasons most prone to radiation fog. In conditions of perfect calm, the fog is often no more than a few feet thick, but light turbulence will transport the cooled air upwards and promote a thicker fog layer. The fog usually evaporates fairly rapidly after sunrise, once incoming radiation begins to warm up the ground.

Advection fog forms when moist air is blown (advected) over a cold surface and is chilled by contact. Typically this occurs over sea areas in early summer when warm winds blowing polewards from tropical areas pass over the cool waters of higher latitudes, creating a sea fog. Areas susceptible to widespread sea fog of this type include the Grand Banks off Newfoundland, and sea areas around north-west Europe. Californian coastal fogs are also similar. With fairly light winds the fog forms close to the water surface, but stronger turbulence may lift the condensed layer to form a low stratus sheet.

Steam fog is generally more localised than the two main types just described. It is likely to develop in situations where cold air blows over much warmer waters. Evaporation from the water body quickly saturates the cold air and the resulting condensation is seen as steaming. The best examples occur in polar areas, where a great contrast in temperature can exist between the very cold air that blows off ice-covered land and the relatively warm surrounding seas. In North Polar regions such fog is sometimes called **Arctic Sea Smoke.**

Precipitation

One of the great puzzles of meteorology has for long been why some apparently similar clouds give rain and others do not. This matter is still not entirely satisfactorily resolved, but some progress has been made towards understanding how precipitation forms within clouds. It is important to realise that there is a large difference between the tiny droplets that make up clouds and the much larger drops which fall as rain. By some means cloud droplets apparently become bigger, but modern studies have shown that they do not grow larger simply by normal condensation processes, as this tends to produce only small droplets of restricted size.

One theory that commands a large measure of support is that formulated by the Norwegian meteorologist Bergeron. The theory suggests that the production of raindrops depends on the coexistence of both water and ice in clouds at temperatures well below freezing. In fact, clouds often do contain supercooled water droplets, down to temperatures as low as $-40°C$, because of the rarity of special **freezing nuclei** that are required to form ice crystals. At these low temperatures, as we have noted earlier in this chapter (Fig. 11.1), it requires more water vapour to saturate the air in respect of a water surface than it does for an ice surface. Thus in a cloud where water droplets and ice exist together, if the air is saturated with respect to the water droplets, it is oversaturated with respect to the ice crystals. The result is that condensation takes place on the ice crystals to reduce the vapour pressure, but spontaneously evaporation commences from the water droplets to restore it. The ice crystals thus grow larger at the expense of the droplets, and eventually become sizable enough to overcome atmospheric friction and cloud updraughts, and fall. On the way down crystals will coalesce to form snowflakes, but these usually melt into raindrops as they pass into warm air layers nearer the ground.

According to this theory, rain comes from clouds which extend well above the freezing level, where the coexistence of water and ice is possible; clouds which are too 'warm' will not produce rain by this mechanism. General observations confirm the essentials of the **Bergeron mechanism**, and the theory has been successfully used as a basis for artificial rainmaking in various countries. Clouds are seeded with dry ice or similar substances which act as freezing nuclei (Chapter Twenty-Four).

Undoubtedly the above mechanism is an extremely important rainmaking process, especially in extra-tropical areas. However, it is likely that there are other mechanisms, since it has been observed that rain sometimes comes from clouds which have great vertical development, but do not reach the temperature levels required by the Bergeron theory. Precisely how these other raindrop-producing processes operate is open to debate, but among the theories suggested are: condensation on extra-large hygroscopic nuclei; coalescence by sweeping, whereby a falling droplet sweeps up others in its path; and growth by electrical attraction between droplets.

Types of Precipitation

The Bergeron mechanism is essentially a **snow**-making process, and provided temperature levels near the ground are below or close to freezing, precipitation will take this form. **Rain** or **drizzle** (small droplets) occurs when temperatures are above freezing, and **sleet** is simply partially melted snow. (In North America sleet sometimes refers to pellets of ice produced by the freezing of rain near the ground.)

Hail appears to be of rather different origin. A hailstone is composed of alternate concentric rings of clear and opaque ice, and is formed by being carried up and down in the vertical currents of a large cumulonimbus cloud. Freezing and partial melting may occur several times before the pellet has become large enough to escape from the cloud.

Thunderstorms

Thunderstorms are a spectacular example of rapid cloud formation and heavy precipitation in unstable air conditions. A prerequisite is that absolute or conditional instability should extend to great heights. This allows powerful updraughts to develop within towering cumulonimbus clouds, which are always associated with thunderstorms. The largest storms tend to occur in tropical or warm regions where the air can hold considerable amounts of water, and they are rarer in polar regions.

A storm frequently consists of several convective cells, each characterised by an updraught. These thunder cells go through a life-cycle which is generally recognised as having three distinctive stages (Fig. 11.8).

In the **developing stage** the initial updraught is formed in response to the uplifting mechanism. The draught is considerably accelerated by the energy provided by the release of latent heat as condensation occurs, and the whole cloud rapidly becomes completely out of thermal equilibrium with its surroundings. Precipitation processes come into operation as the cloud develops beyond the freezing level, but the great strength of the updraughts initially prevents the snow and ice from falling.

The **mature stage** is reached with the sudden onset of heavy rain, perhaps accompanied by thunder and lightning. The precipitation drags cold air down

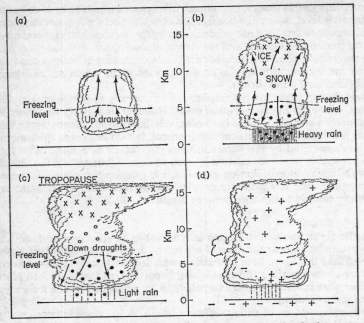

Fig. 11.8. Stages in the life-cycle of a thunderstorm: (a) developing stage;
(b) mature stage; (c) dissipating stage; (d) distribution of electrostatic
charges in latter stages.

with it, and this is often felt on the ground immediately prior to the rain.
During the latter part of the mature stage the upper parts of the cloud, which
are composed of ice particles, reach the tropopause. Here the cloud spreads
out into a characteristic anvil shape, which may be elongated by high altitude
winds.

As the supply of moisture in the cell is gradually exhausted, the energy
slackens and the storm passes into the **dissipating stage.** This is characterised
by downdraughts which spread out below the cloud, thus preventing any
further convective instability in the immediate vicinity. However, new cells
may be initiated by the meeting of cold downdraughts from cells some kilometres apart, triggering off the rise of warm air in between.

Lightning

Fig. 11.8c shows the distribution of electrical charges in a thundercloud and
in the adjacent ground area. Lightning occurs to relieve the electrical tension
between oppositely charged areas: this may be between the cloud and the
ground, or within the cloud. Broadly speaking, the upper part of the cloud is
positively charged, and the lower part negatively, except for a small positively
charged region around the rain area.

Modern views on the origin of this pattern of charges are that they are based

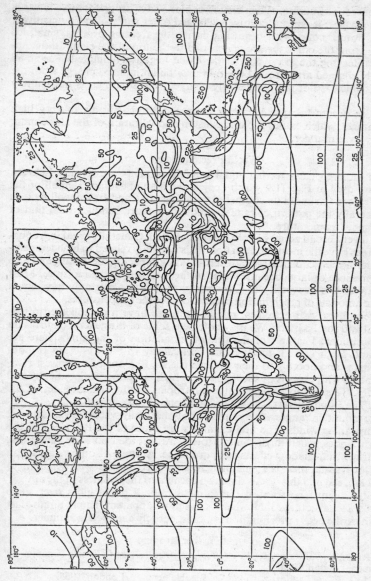

Fig. 11.9. World distribution of precipitation (figures in cm).

on the effects of freezing. When a supercooled water droplet freezes inwards from its surface, it gains a positively charged outer surface and a negatively charged 'warm' core. With rupturing during the freezing process, small ice splinters from the outer surface are carried to the top of the cloud in the up-draughts, leaving the lower part of the cloud negatively charged. The small positively charged area around the rain may be an induced charge from the ground, which is normally negatively charged, but its origin is not entirely certain.

Thunder occurs because lightning heats the immediate air to very high temperatures, which causes rapid expansion and vibration of the air column, heard by us as thunder.

World Precipitation

Lines joining places having equal rainfall amounts are termed **isohyets**. These are used in Fig. 11.9 which depicts the annual average rainfall of the world. This map should be compared with Figs. 10.6 and 10.7; it will be seen that generally the pattern of rainfall amounts is more complex than that of mean annual temperature.

Only a very broad zonal pattern can be detected. Generally speaking, equatorial areas have the most precipitation. This is mainly related to high temperatures and the consequent large moisture-holding capacities of the air, and also to the presence of large oceanic water surfaces to supply the moisture. Most of the rainfall is convectional. Conversely, polar areas can only have small absolute amounts of precipitation because of cold air temperatures.

Between these two extremes, middle latitudes have a complicated distributional and total pattern. We may note here some of the general controls on these variations. Features of the general circulation of the atmosphere are important. High rainfall totals relate particularly to the westerlies in both hemispheres and notably to their cyclone tracks. Regions of the lowest rainfall coincide with regions of subsiding air, which is warmed adiabatically and made dry. This occurs mainly in the subtropics on the eastern sides of the oceans. The Sahara desert is an extension of one of these areas.

Large mountain ranges affect rainfall totals markedly where they lie athwart moist prevailing winds. The best examples are those of the Rockies and the southern Andes, where high rainfall totals are recorded on windward sides, and marked rain-shadow effects exist in the lee. Altitude also plays a part on a more local scale: there is a general increase of precipitation with height up to about 2 km. Beyond this, totals diminish because of the coolness of the air.

Finally, it must be stressed that annual rainfall totals gloss over some marked seasonal contrasts. Nowhere is this more apparent than in South-East Asia, where the heavy precipitation is almost entirely confined to summer.

Suggested Further Reading

Barry, R. G., and Chorley, R. J., *Atmosphere, Weather, and Climate* (Chapter 2), Methuen, London, 1970.

Battan, L. J., *Cloud Physics and Cloud Seeding*, Heinemann, London, 1965.

Ludlam, F. M., and Scorer, R. S., *Cloud Study. A Pictorial Guide*, Murray, London, 1966.

Mason, B. J., *Clouds, Rain and Rainmaking*, Cambridge University Press, London and Cambridge, 1962.

Simons, M., 'Atmospheric convection', *Geography*, 1970, **551**, 196–203.

CHAPTER TWELVE

AIR MOTION

Vertical air movement of the type described in the previous chapter is primarily important in a local context. On the other hand, horizontal air movement, which we commonly call 'wind', is of significance at many scales, ranging in size from small eddies to major hemispherical wind systems, and also tends to be much more powerful than vertical motion.

The basic impulsion to air movement on a large scale lies in the unequal global energy budget discussed in Chapter Ten. One of the major equalising factors in this budget is the transfer of latent or sensible heat from one place to another by bodily air movement. The detailed effects of this on the global circulation will be looked at in the next chapter. The point of concern here, however, is that the variable heating of different parts of the atmosphere sets up variations in pressure, which in turn sets the air in motion. There is, in fact, an intimate relationship between winds and pressure, and a knowledge of pressure variations is a prerequisite to understanding air motion.

Pressure Variations

Pressure is normally measured in millibars and spatial variations of pressure are depicted on maps by means of isobars, which are lines connecting places having the same barometric pressure. To the practising meteorologist, pressure conditions are an extremely important aid in understanding and predicting weather situations. Thus to facilitate comparison on a pressure map between one weather station and the next, pressure readings are usually adjusted to their sea-level equivalents: this is a calculation of what the pressure would be at that particular time if the station were at sea-level. This device effectively eliminates the consideration of elevation as a pressure factor, and allows horizontal trends and changes of pressure to be seen more clearly.

The mean sea-level pressure over the globe is about 1013 mb. The range of normal variation can extend up to about 1060 (in the Siberian winter anticyclone) and down to 940 mb, although the latter figure may be temporarily exceeded in tropical cyclones. At any one locality the trend of the change of pressure with time usually has more weather connotations than the absolute reading itself. Hence falling pressure generally heralds the onset of poorer weather, and a rising barometer suggests a trend towards brighter conditions.

Pressure Gradients

The gradual change of pressure between different areas is known as the **barometric slope** or, more commonly, as the pressure gradient. Certain analogies may be made here with the contour features of a topographic map. The direction of steepest gradient is at right angles to the isobars. The closer the isobars are together, the greater the pressure gradient; for example, widely spaced isobars indicate a weak pressure gradient. The terminology of some of the features of a pressure map, such as 'ridge' and 'col', which are illustrated

135

in Chapter Fourteen, is also used in similar fashion to that of a contour map. The terms 'high pressure' and 'low pressure' do not usually signify any particular absolute values, but are used relatively.

Surface Pressure Belts

Sea-level pressure conditions over the globe for both January and July (Fig. 12.1) show some marked differences between the two hemispheres. The northern hemisphere tends to have the greater seasonal contrasts in its pressure distributions, and the southern hemisphere exhibits much simpler average pressure patterns overall. These differences are largely related to the unequal distribution of land and sea between the two hemispheres. Ocean areas, which dominate the southern hemisphere, tend to be much more equable than continents in both temperature and pressure variations.

The most permanent features of both hemispheres are the subtropical high-pressure belts. These are particularly evident and uniform over ocean areas, and thus in the southern hemisphere around latitude 30°S there is an almost continuous belt of high pressure, which is only broken by small low-pressure areas in summer over Australia and South Africa. The average pressure in this belt generally exceeds 1026 mb. In the northern hemisphere, the corresponding belt at 30°N is made more discontinuous by the presence of land masses, and high pressure usually occurs only over the ocean areas as discrete cells; these are termed the **Azores** and **Hawaiian cells** in the Atlantic and Pacific areas respectively. Over the continental areas at this latitude—namely the southwest United States, the Sahara and southern Asia—major seasonal fluctuations are apparent: high-pressure areas exist in winter, but these are replaced in summer by lows which are largely induced by overheating of the ground.

Equatorwards of the subtropical belts, pressure gradually falls towards the **equatorial trough**, where pressure values are normally of the order of 1008 to 1010 mb. The trough is coincident with the zone of maximum insolation. In the northern hemisphere summer, it lies well to the north of the equator, especially over the continental areas, reaching as much as 25°N over the Indian subcontinent. In January the trough lies just to the south of the equator; it appears that the landmasses of the southern hemisphere are not of sufficient size to cause a significant southward displacement. Over sea areas, the region of the equatorial trough is sometimes called the **doldrums**, noted for its slack pressure gradients and becalming effect on sailing ships.

In temperate latitudes, pressure is generally lower than in the subtropical areas. Perhaps the most notable feature of these middle latitudes is their moving depressions and pressure ridges, which are not apparent on maps of mean pressure. A gradual decrease of mean pressure in these regions, reaching a minimum at about 60° of latitude, is best seen in the southern hemisphere, where the belt is referred to as the **sub-antarctic low**. In the northern hemisphere, the situation is slightly more complicated. Strong winter low-pressure cells exist in the Icelandic and Aleutian areas, but over Siberia and Canada the coldness of the land surface chills the air to cause the development of large high-pressure cells over these regions. In summer, thermal heating over the continental areas replaces the winter highs with weak low pressure, and in July a general belt of relatively low pressure exists right round the northern hemisphere at approximately 60°N (**sub-arctic low**).

In the polar areas of both hemispheres, pressure tends to be relatively high at all seasons. This is more marked over the land area of the Antarctic continent than over the ocean areas of the North Pole.

Forces Governing Air Movement

The existence of pressure differentials in the atmosphere is the immediate primary force causing air movement. The **pressure gradient force** always acts down the pressure gradient, attempting to cause the general movement of air away from high-pressure towards low-pressure areas. The force exerted is proportional to the steepness of the gradient. This should mean that winds ought to blow at right angles to the isobars, but in practice wind hardly ever flows solely under the influence of the pressure gradient force; in fact it tends to flow more nearly parallel to the isobars. This is because there are other important controls on air motion. These principally include: the **Coriolis force**, which is the deflecting force caused by the Earth's rotation; **centripetal force**, which acts around circulatory pressure systems; and the **frictional force** exerted by the Earth's surface.

Coriolis Force

This is named after the French physicist Coriolis, who in the nineteenth century formalised the concept of the Earth's deflecting force. The effect of the rotation of the Earth about its axis is to cause an apparent deflection of moving air from its original path. This deflection is always to the right of the direction of motion in the northern hemisphere, and to the left in the southern, whatever the original bearing of the wind. The phenomenon affects all freely moving objects, including air, ocean currents, rockets and projectiles. Although its effects seem real enough to anyone on the ground, the force is sometimes called 'apparent', because if viewed from outer space, objects moving across the face of the Earth would not in fact appear to be deflected. In relation to star positions, they would travel in a straight line, while the Earth rotates beneath them. An observer on the ground, however, is naturally unaware that he himself is turning with the Earth and has a different view of things, and thinks that it is the moving objects that are being deflected. With regard to our consideration of air movement, it is simplest to accept that the deflection of winds is being caused by an actual 'force', even though this is not strictly correct.

The degree of the deflecting force varies with the speed of the moving air and with latitude. The faster the wind, the more ground it covers in a given time, and the greater the effect of rotation can be. Near the equator, where the Earth's surface is spinning round in a plane almost parallel to the axis of rotation, the Coriolis force is very slight. In higher latitudes, however, it has marked effects.

The Geostrophic Wind

Neither the pressure gradient force nor the Coriolis force can account for the fact that global winds generally blow with steady force and direction. If the pressure gradient were to act alone, it would cause winds to blow directly down the gradient with increasing acceleration; if the Coriolis force were paramount, winds would in theory eventually circumscribe a circle. In reality wind directions adopt a condition of equilibrium or balance between various forces, the most important of which is the **strophic balance** which exists

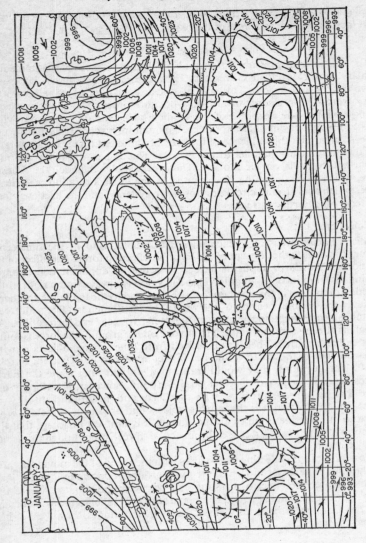

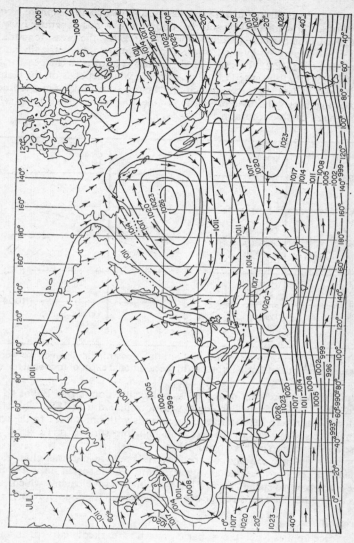

Fig. 12.1. Average barometric pressures and winds for January (top) and July (bottom) (figures in mb).

between the pressure gradient and Coriolis force. In the free atmosphere, above the level of air flow affected by surface topography, winds generally blow at right angles to the pressure gradient: this indicates that the pressure gradient force is exactly balanced by the Coriolis force acting in a diametrically opposite direction (Fig. 12.2). This sort of air motion is known as the **geostrophic wind**. It follows that the speed of the geostrophic wind depends on the factors that govern its balancing forces, principally the pressure gradient and latitude, and knowing these, the meteorologist is able to predict geostrophic wind forces with reasonable accuracy.

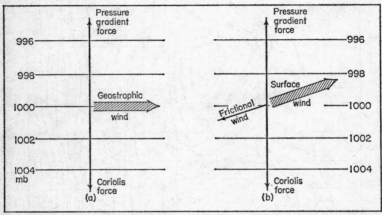

Fig. 12.2. Forces governing air motion: (a) geostrophic balance between pressure gradient and Coriolis forces; (b) the additional effect of frictional force.

A qualitative expression of the geostrophic situation is **Buy Ballot's Law**, named after the Dutch meteorologist who formulated it in 1857. This states that, for the northern hemisphere, if one stands with one's back to the wind, low pressure always lies to the left, and high pressure to one's right. The reverse applies in the southern hemisphere.

Not all winds are exactly geostrophic. As pressure patterns change, the balance is upset, but the wind always strives to readjust itself until it obtains the new geostrophic speed. Two of the most important elements which cause winds to be other than geostrophic are centripetal and frictional forces.

Centripetal Force

This force applies when the isobaric pattern is markedly curved, as within cyclonic systems or around high-pressure centres. The fact that air is following a curved path means that in addition to the pressure gradient and the Coriolis force, a third force is acting centripetally, pulling the air inwards. Wind which is in balance with these three forces is known as the **gradient wind**.

Fig. 12.3 illustrates the gradient wind situation for high- and low-pressure areas. Motion around a low-pressure area, anticlockwise in the northern hemisphere, is termed cyclonic, and in this case the result of the centripetal effect is to make the Coriolis force weaker than the pressure gradient force;

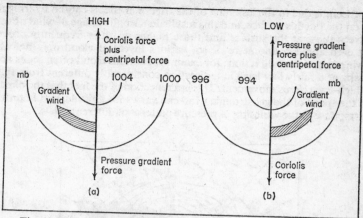

Fig. 12.3. Forces acting on wind flow around (a) high-pressure and (b) low-pressure systems.

the wind is 'subgeostrophic'. The anticyclonic flow in the high-pressure case, clockwise, is 'supergeostrophic', since the Coriolis force exceeds the pressure gradient force.

Frictional Forces

In the lowest parts of the atmosphere, normally below 750 m, the frictional drag exerted by the ground on the airflow above has an effect on the balance of the other wind forces. Friction lessens the speed of the wind, and in doing so weakens the Coriolis force. This allows the pressure gradient to assert its greater strength by causing the air to flow more towards low pressure. Thus the usual situation is that surface winds flow at a slight angle to the isobars (see Fig. 12.2b). This may also be seen in examples of synoptic charts in Chapter Fourteen. The deviation from the geostrophic wind direction which obtains at higher levels is generally of the order of 10 to 20 degrees. The deviation naturally tends to be less over smooth surfaces, notably the oceans. It also tends to be less in strong wind flow, and greatest in calm conditions at night.

Upper Air Motion

The study of pressure systems and the behaviour of winds in the higher parts of the troposphere has assumed great importance since the Second World War. A great deal of information about upper air conditions has been gathered from aircraft, satellites, and particularly from radio-sonde balloons. On practical grounds a knowledge of upper winds is essential to modern aircraft navigation. As far as meteorology is concerned, it is now realised that the causes of weather on the ground are intricately bound up with what happens at higher levels. This applies especially to the development of anticyclones and depressions and to the general circulation of winds around the globe. Such phenomena can only be appreciated as three-dimensional features.

Broadly speaking, wind speeds tend to increase with altitude because of lower air densities. In the first kilometre of height, part of this increase, as we have seen, is due to the lessening of the frictional effects caused by irregularities on the Earth's surface, and this results in a small change of wind direction between those at the surface and those higher up. It is frequently observed that even above a kilometre or so, various layers of cloud may not all be moving in the same direction: for example, cumulus cloud often scuds across the sky at a fairly low level in a direction considerably different from that of the high-level cirrus movement. This is a reflection of the fact that the direction and the speed of the geostrophic wind can vary with altitude, and in turn this is a response to the variation of pressure patterns at different levels.

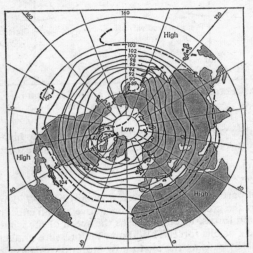

Fig. 12.4. Mean contours (in hundreds of feet) of the 700 mb pressure surface in January in the northern hemisphere.

Upper Pressure Patterns

There are several ways of depicting spatial variations of pressure at high level. One is to use a **constant-level chart**; a particular height above sea-level is selected—say 2 km—and the isobars are drawn in exactly the same way as on a normal sea-level chart. A more common practice, however, is to take a fixed atmospheric pressure—say 500 mb—and to plot its height above sea-level at various points. This is known as a **constant-pressure chart** (Fig. 12.4). On this, the normal pressure terminology of 'highs' and 'lows' can still be used, but reference is of course to height, not pressure. A third method of depiction is the **thickness chart**. This is described below in connection with thermal winds.

One notable finding of upper atmosphere exploration is that pressure systems observed on the ground do not necessarily persist to great height. Cold anticyclones, for example, which at ground level are often characterised by intense high pressure, seldom persist in height beyond 2 km, and are replaced by relatively warm air with a lower pressure than surrounding regions.

In contrast, warm anticyclones frequently intensify their relatively higher pressure with height.

On a global scale, pressure patterns higher up tend to be much simpler than those at the surface level, largely because of the diminished thermal and mechanical effects of land masses. The average situation for one month in the northern hemisphere (Fig. 12.4) reveals a generally falling pressure gradient from the subtropical areas towards the poles. The gradient is strongest in winter, when the temperature contrasts between the respective polar areas and the equator are most marked. To some extent, high-pressure cells still exist at height over the subtropical regions, and as such these are the only areas where the surface pattern substantially persists to higher levels. It is thought, therefore, that these latitudes have a key role to play in the general circulation of the atmosphere. Elsewhere over the globe, surface pressure features are poorly reflected, or reversed.

Thermal Winds and Jet Streams

Changes in pressure distributions with height are largely related to changes of temperature. We can see how this can be so with reference to two adjacent columns of air in the troposphere depicted in Fig. 12.5. At ground level the pressure exerted by the two is the same, but important changes ensue if we assume that column A is warmer, and therefore less dense throughout than B. This means that for any level higher up in the two columns, for instance at 2 km, there is a greater pressure of air still above this level in column A than in column B. In other words, a pressure gradient from A to B gradually develops and intensifies with height, where none existed at the surface. We can thus visualise a gradual change of velocity of the geostrophic wind with height, the wind at the top of the air layers being very much stronger than that lower down. Such a change of wind velocity is known as **wind shear**. There is also in practice likely to be a gradual change of wind direction.

It is possible to calculate fairly easily on a **vector diagram** the mean wind

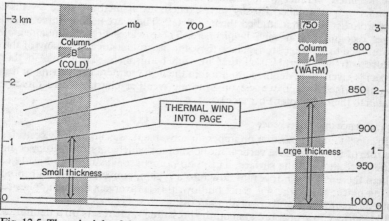

Fig. 12.5. The principle of thermal winds: different vertical temperature gradients in the two columns create an increasing pressure gradient.

direction and force between the geostrophic winds at the top and bottom of a layer. This vector difference is known as the **thermal wind**, since it derives from temperature contrasts in the atmosphere. It is a useful concept in meteorology since it can be used to predict the influx of warm or cold air into a region.

Fig. 12.5 illustrates that the **thickness** of the atmosphere between any two pressure levels is related to temperature. Warm air causes a large thickness, and a small thickness results from cold air. This is the basis for the construction of a thickness chart, which is another method of plotting upper air patterns. The usual thickness chart employed in meteorology is for the layer 1000–500 mb. On a thickness chart, the thermal wind blows at right angles to the thickness lines, which are in effect isotherms. In the northern hemisphere, where the Coriolis deflection is to the right, cold air always lies to the left of the thermal wind when viewed downwind, and to the right in the southern hemisphere.

Applying this on a global scale, the gradual poleward decrease of temperature in the atmosphere from the equator should result in a large westerly component in the upper winds, and in fact we find that in both hemispheres most upper winds are westerly.

High-flying aircraft during the Second World War from time to time encountered upper winds of very great velocity. These have since been the subject of considerable study and are now known to be concentrated bands of rapid air movement, which are termed **jet streams**. They occur near the top of the troposphere, and are in reality intense thermal winds, being associated with latitudes where the poleward temperature gradient is particularly strong. Two such zones occur in each hemisphere. One, the subtropical jet stream, occurs at about 30° of latitude, and the other, the polar front jet stream, is associated with the polar front zone in each hemisphere (see Fig. 13.2c).

Wave Motion in the Upper Winds

The flow of the geostrophic winds around the globe is not in the form of straight lines, parallel with latitude, but is rather in huge serpentine or meandering paths called **Rossby waves**. These are named after the Swedish meteorologist who first studied them in detail. There are usually three to six waves recognisable in each hemisphere. These waves are of fundamental importance to weather study, since they appear to link together many of the features of the general circulation in each hemisphere. Their role in this respect is given more detailed attention in the next chapter. The waving of the frontal surfaces which we see on synoptic weather charts is also directly related to them (Chapter Fourteen).

Convergence and Divergence

In concluding this section on upper air movement, we can usefully consider the role of the large-scale vertical motions of air which connect the upper air patterns to those at the surface. The main points are illustrated in Fig. 12.6. Where there is a net horizontal flow into a region, convergence is said to be taking place; where there is a net outflow, this is divergence. This applies to both surface systems and upper air patterns.

In a surface low-pressure system, the pressure gradient sets up a net inflow of air. The actual wind motion, as stated earlier in this chapter, is cyclonic, blowing at a slight angle to the isobars. However, to maintain the low-pressure system, this influx of air has to be relieved by vertical ascent, otherwise the depression would rapidly fill up and disappear. In turn, the vertical movement of air is itself accompanied by outward divergence at high level. Similarly, in a surface high-pressure system, there is a divergence of air at low levels which is maintained by convergence higher up. In either situation, the changeover from convergence to divergence in practice usually occurs at about the 600 mb level in the atmosphere.

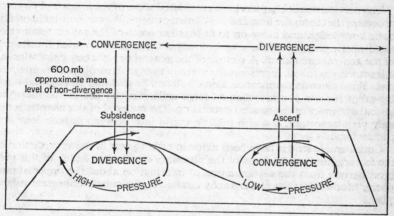

Fig. 12.6. Vertical movements of air involving mass convergence and divergence in adjacent high- and low-pressure systems.

Marked convergence and divergence in the upper atmosphere usually only occur when the geostrophic Rossby wave pattern is weak or has broken down (see Fig. 13.4c). The vertical motions involved are usually very slow compared with the horizontal components. However, vertical transport of air can be very persistent and may last over a period of several days. These air movements go some way towards explaining why many anticyclones, characterised by descending and therefore warming air, are relatively cloudless, whereas in low-pressure regions the ascent of air is liable to give cloud and bad weather.

Suggested Further Reading

Barry, R. G., and Chorley, R. J., *Atmosphere, Weather, and Climate* (Chapter 3), Methuen, London, 1970.

Newell, R. E., 'The circulation of the upper atmosphere', *Scientific American*, Vol. 210, pp. 62–74, 1964.

Peterson, J., *Introduction to Meteorology* (Chapters 9 and 10, 3rd edn.), McGraw-Hill, New York, 1969.

Reiter, E. R., *Jet streams*, Doubleday, New York, 1967.

Riehl, H., *Introduction to the Atmosphere* (Chapter 6, 2nd edn.), McGraw-Hill, New York, 1972.

CHAPTER THIRTEEN

THE GENERAL CIRCULATION

The previous three chapters have considered some of the important physical processes that form the foundations of meteorology. We can now build on this basic knowledge, and move on to fit together some of the salient features of atmospheric energy, moisture and movement into the integrated framework of the general circulation. A picture of the general or **planetary circulation**, as it is sometimes called, represents the mean of many meteorological events, and gives them climatic significance. Equally, it can be said that an essential starting-point for the consideration of large-scale climates lies in the meteorological dynamics of the general circulation. The material of this chapter is not only very important, but also fulfils a useful link between meteorology and climatology.

Considerable strides have been made in recent years in the appreciation of the features and mechanisms of the planetary circulation. Much of this progress derives from the accumulation of information about the upper atmosphere. Modern satellite photography has had a particularly important role to play.

The Planetary Wind Belts

Broadly speaking, all major wind systems around the globe, whether at the surface or in the upper atmosphere, can be regarded as predominantly zonal (latitudinal) in character. However, applying the principles elucidated in the previous chapter, we may expect upper winds to be more truly zonal than those near the surface, which, largely as a result of the effect of ground friction, have significant north–south components.

Surface Wind Belts

In each hemisphere, two wind belts effectively dominate the circulation (see Fig. 13.1). One of these is the **trade-wind belt**, blowing from an easterly quarter. These winds blow over nearly half the surface of the globe, between latitudes 30° north and south. The permanency of the subtropical high-pressure zones (see Fig. 12.1) has an important bearing on the well-known constancy of the trade winds which emanate from the equatorward side of the high-pressure cells. The two trade-wind systems converge towards each other in the equatorial trough. At certain times of the year there exists a narrow zone of **equatorial westerlies** blowing between the trade winds. These westerlies occur in summer in each hemisphere, when the trade-wind belt has moved polewards, and are particularly noticeable over the low-latitude land areas of the northern hemisphere. The south-west monsoon of India is an exaggerated expression of these winds.

The second major surface-wind belt in each hemisphere is the mid-latitude **westerlies**, which develop out of the poleward sides of the subtropical high-pressure cells. The westerlies of the southern hemisphere are the stronger and the more persistent, and locally carry evocative names such as 'the Roaring Forties'. Interference to the westerly flow by land areas is minimal here, and ocean weather stations in the belt have a 70 to 80 per cent preponderance of westerly winds. However, the corresponding wind regime of the northern hemisphere has winds which are much more variable in force and direction. The irregularity of the surface relief, and the fluctuations of local pressure patterns over the land areas, both tend to break up the westerly flow. Many stations here have winds blowing from all points of the compass, with westerly winds perhaps only just predominating, and it is rare for a station to have winds from a westerly bearing on more than half the days in the year.

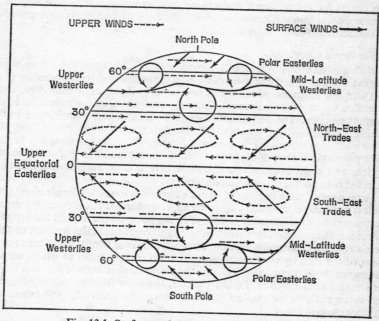

Fig. 13.1. Surface and upper planetary wind belts.

Polewards of the main westerly belt in each hemisphere, high-latitude areas are generally regarded as being in the regime of the **polar easterlies**. These winds tend to be most in evidence on the poleward sides of transitory depressions, but elsewhere easterly winds are rather less obvious. In the Arctic, this is partly because the source of these winds, the polar cap high-pressure area, is only a seasonal phenomenon. In Antarctica, where the polar anticyclone is a more permanent feature, easterly winds appear to be more reliable.

Another way of looking at the surface circulation pattern is to recognise the main zones of convergence and divergence, which are of course fundamental to the maintenance of the system. Divergent air flows outwards from surface high pressure, and the most important regions of divergence on the globe are the subtropical high-pressure cells. These are areas of relatively calm winds, sometimes given the name 'Horse Latitudes', for reasons now rather obscure. Other regions of weaker divergence include the seasonal anticyclones over Asia and North America. Three major zones in which convergence markedly occurs encircle the globe. One of these zones lies in the equatorial trough between the inblowing trade winds and is termed the **Inter-tropical Convergence Zone (ITCZ)**. Along its length, this is a rather variable feature, being easily recognisable only where there are noticeable temperature and humidity contrasts in the two converging trade winds; elsewhere a doldrum situation may prevail. The other zones of air mass convergence are the polar front zones of each hemisphere, existing as major synoptic features between the westerlies belt and the polar easterlies. On a chart of mean wind-flow (streamline chart) for the northern hemisphere, convergence in this zone is usually most apparent in the Aleutian and Icelandic lows, but general convergence of air occurs in all frontal situations.

Upper Zonal Circulation

With the loss of north–south components of direction, the upper winds around the globe become essentially true easterlies and true westerlies (Fig. 13.1). In low latitudes, upper winds appear to be dominantly **easterlies**. Their extension varies with the seasons, depending primarily on the migration of the thermal equator. The belt encroaches polewards in each hemisphere in the summer, reaching as far as about 20°N in July, but at other times of the year these winds are more limited in extent. A marked tropical **jet stream** is known to be present in the upper easterlies when they are over India and north-central Africa, and a probable southern hemisphere counterpart to this has been detected over northern Australia in January.

From about fifteen degrees of latitude almost to the poles the air circulation aloft is generally dominated all year by **upper westerlies**. These steadily increase in speed polewards, forming one large **circumpolar vortex** in each hemisphere. These winds are sometimes referred to as the **zonal westerlies** because of their persistent character and direction. However, in these winds, significant variations of the Rossby wave pattern occur from time to time, to which we will subsequently give our attention. The upper westerlies include the two major jet streams referred to in the last chapter, the **polar front** or **primary jet**, and the more variable **subtropical jet**, both normally occurring just below the tropopause (see Fig. 13.2).

The Mechanics of the Circulation

In seeking to explain how this pattern of planetary circulation comes about, and how changes in it may possibly be anticipated, it may be recalled that the fundamental drive behind the atmospheric system is the imbalance of heat energy between the equator and the poles, which the atmosphere attempts to correct by poleward heat transport. In its simplest form, this should operate like a gigantic heat engine and produce a **single-cell** circulation in each hemisphere, in which there is an outflow of warm air at high level from the equator towards the poles, and a return surface flow in the opposite direction (Fig. 13.2a). In this theoretical model for the northern hemisphere, the Earth's rotational force would be expected to cause the upper flow to be predominantly westerly, and taking into account the additional effects of friction, the surface flow to be between south and east. However, although this model illustrates certain broad principles, we can immediately recognise that it is rather at variance with the observed pattern of global winds.

As long ago as the eighteenth century, a more sophisticated **tri-cellular** model was proposed (Fig. 13.2b) to fit in with the then known facts of the surface wind system. Hadley, in 1735, argued logically that in each hemisphere there must be a low-latitude cell which relies for its driving mechanism on the rising of heated air at the equator, its outflow aloft polewards, and sinking at higher latitudes before returning to the equator. This we call nowadays a **thermally direct** cell. Similarly, in this three-cell scheme, another thermally direct cell has been thought to exist at the pole, the contraction of the air column in these cold regions initiating inward flow at high level and outward

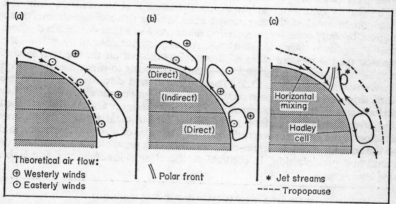

Fig. 13.2. Models of the general circulation: (a) simple thermally direct model; (b) three-cell model; (c) Palmen's model.

divergence at low level. The middle, or mid-latitude, cell is considered to be thermally indirect, largely maintained by the circulations in the other two. This tri-cellular concept has remained the basis for understanding the general circulation until quite recently. It fits in reasonably well with the surface winds, but with increasing information about the upper winds, it is now realised that this model is inadequate. In particular, it does not account for the great preponderance of westerly winds at high level in middle latitudes.

The model in Fig. 13.2c, first suggested by Palmen in 1951, is a good example of modern thinking about the circulation system. Although not all details are yet satisfactorily understood, the broad pattern suggested by this model is thought to be correct.

Perhaps one of the most important advances in recent times has been that the mode of heat transport from equator to poles does not have to be explained solely in terms of air movement in the vertical plane, as these various models would tend to suggest. It is increasingly appreciated that a great deal of heat exchange is also accomplished in a horizontal sense, particularly in middle latitudes, where large masses of air penetrate north and south past each other in wave and frontal patterns. This means in effect that **frontal systems** are a major force in the maintenance of the general circulation, and not merely transitory features superimposed on it. With this important point in mind we can now look at the workings of the principal parts of the modern circulation model.

Low-Latitude Circulation

It is nowadays thought that only in the tropics is it likely that regular cell-like arrangements exist as a fundamental method of circulation. These, appropriately enough, are called **Hadley cells**, and their broad mechanism is probably as Hadley first suggested. But on the evidence of some of the weather patterns produced in the tropics, this circulation is not as simple in detail as it first looks. In order to gain a truer picture of reality, we have to add to our cross-sectional notion of the Hadley cell the plan arrangement of the several high-pressure cells of the subtropics. Although it is rather difficult without a three-dimensional model to visualise exactly what this means in terms of air motion, broadly speaking, circulation takes place in a series of inclined planes tilted upwards at their poleward ends. Air moves upwards and away from the equator on the western sides of these cells—that is, on the western side of oceans in the northern hemisphere—and downwards towards the equator on their eastern limbs (Fig. 13.3). These upward and downward motions are fairly slow and superimposed on the trade winds and upper westerlies, but they do have several major consequences. The trade winds are alternately weakened and strengthened in various parts of the intertropical region; fronts may develop in the areas between adjacent cells because of air-mass contrasts; and at the equator itself, the areas of upward-moving unstable air are denoted by cloud and precipitation, in contrast to the clear skies characterising stable descending air.

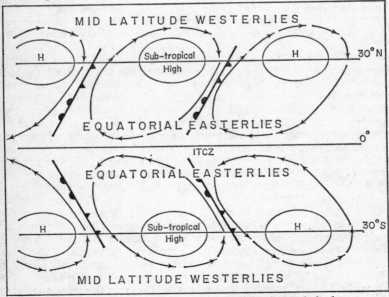

Fig. 13.3. Plan view of the general circulation in low latitudes.

Another factor of major significance in the operation of the tropical circulation is its wholesale migration with the seasons. Although the ITCZ and the axis of the equatorial trough are not always necessarily coincident, both follow the migration of the overhead sun, normally lagging about six weeks behind it. The greatest poleward displacement of the ITCZ occurs in the northern hemisphere, where it reaches as much as 25°N over the Indian sub-continent. This and other strong migrations are encouraged by the development of summer low-pressure in South-East Asia, Africa and Australia. The movement polewards of the **thermal equator** allows the Hadley cell circulation of the opposite hemisphere to encroach over the equator. On doing so, the wind directions associated with this cell are liable to be reversed because of the different effect of the Coriolis force in the other hemisphere. This is one possible explanation of the surface equatorial westerlies and the upper easterlies.

The migration of the Hadley cell system also means that the outer subsiding parts of the circulation reach as far as 40°N in parts of the northern hemisphere in August and September, virtually guaranteeing warm sunny weather to the Mediterranean and southern California.

Some authorities think that the ITCZ has a double structure. Instead of there being one zone of general convergence between the two opposing trade-wind systems, which migrates back and forth across the equator, it has recently been suggested that there may be one band of precipitation and cloudiness associated with convergence north of the equator, and one to the south. The apparent motions of the ITCZ across the equator may therefore be connected with changes in the intensity of the two belts rather than actual movement, although partial migration may occur. In between the two convergence belts, westerly surface winds would occur, and aloft, sinking easterly flow would be found, which might be a better explanation of the observed winds. The problem is not yet fully resolved.

Mid- and High-Latitude Circulation

Polewards of the outer parts of the Hadley cells, the mechanisms of the general circulation assume a rather different character. In Palmen's model (Fig. 13.2c) only a very modified form of weak indirect cell exists, if at all. We may note in the model the continuing dominance of westerlies in the upper flow, the inclined polar front, and the presence of horizontal mixing in middle and high latitudes.

If we translate the polar front zone shown here into plan view, it would be characteristically wavy, as commonly revealed on synoptic weather charts, associated with large vortices of air movement or with travelling depressions and anticyclones. It is in these vortices that there is considerable bodily north–south transference of air. The lateral movement of these vortices around the globe at these latitudes gives rise to some characteristic weather sequences, which are dealt with in the next chapter. The occurrence and degree of waving on the polar front is rather variable, but it is known that it is connected with the changing amplitude and wavelength of the Rossby long waves, which thereby have an important control on our weather.

Variations in the upper flow in middle and high latitudes are usually referred to in terms of the **zonal index**, which is a measure of the strength of the upper westerlies. In a fairly average index situation (Fig. 13.4a), three to six Rossby waves encircle the atmosphere in amplitudes covering 15–20° of latitude and up to 60° of longitude. The troughs in the waves reach towards lower latitudes, generally between 35° and 45°N, and this is usually where the jet streams are at their strongest. Although the troughs and ridges in the wave pattern do migrate eastwards around the globe, they normally tend to persist in preferred positions for several weeks at a time, guiding the track of the passing depressions beneath them.

In a **high index** situation, the waves are flat, or hardly recognisable (Fig. 13.4b). The westerly flow is strong, and the whole belt contracts towards the

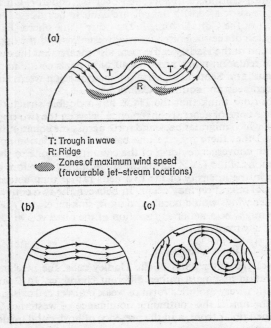

Fig. 13.4. Zonal flow in the upper westerlies: (a) the most commonly observed situation; (b) high zonal index; (c) low zonal index.

pole, usually north of 40°N. In these circumstances, the north–south exchange of energy or heat is minimal, either in the upper winds or at surface level. The effect on the weather is that over most of the area towards the middle latitude margins of this circulation, the climate is mild. In a **low index** period, matters are very different (Fig. 13.4c). The Rossby wave pattern progresses towards this situation with a gradual slowing down in the speed of the circumpolar vortex, an increase in the latitudinal amplitude of the waves, a shortening of the wavelength, and eventually the wave system breaks down into a cellular pattern. This allows strong meridional flow; strong incursions of cold air are enabled to move southwards in the troughs of the cellular pattern, whereas in other places warm air can move a long way northwards. A low index situation is therefore a time of maximum poleward transport of energy in the mid-latitude circulation.

The effects of major variations in the Rossby wave pattern on surface weather are usually most noticeable in winter, when the westerly circulation is at its strongest. A breakdown in the flow can lead to the development of some of our most persistent cold anticyclones, which encourage the penetration of polar air on their eastern margins to relatively low latitudes. However, although weather-forecasters can usually predict the likely consequences of a change in the upper air pattern, the precise causes of the timing of these variations are as yet uncertain.

Surface Features and the Circulation

We have from time to time discussed some of the consequences of the varying character of the Earth's surface on atmospheric phenomena, particularly those of pressure patterns and surface winds. The basic cause of these differences lies in the contrast in heat properties between land and sea. A very good illustration of this on a small scale is the land and sea breeze (Fig. 16.2). In global terms, ocean temperatures vary little through the year, except where seas are shallow, because excessive heating and cooling are absorbed by the mixing of the surface waters with those from below. On the other hand, continental areas obviously do not have this facility and experience a much greater range of temperatures. These effects have a major bearing on the distribution and character of regional climates, which are discussed in Chapter Fifteen. In addition, these differences are also large enough to have some effect on the general circulation.

Differences Between Hemispheres

On a homogeneous globe, we could reasonably expect the planetary circulation to be the same in both northern and southern hemispheres; but on the Earth, the northern hemisphere has a much greater land area than the southern, and this results in certain distinguishing features in the two hemispherical circulations.

First, the general circulation of the southern hemisphere is demonstrably more uniform in terms of latitudinal zones. In the northern hemisphere, the spacing of the oceanic and continental areas inevitably irregularises the circulation, markedly near the surface, less so higher up. On the other hand, in the southern hemisphere there are no large continents to create strong thermal effects on a regional scale, and seasonal contrasts are less in evidence than in the northern hemisphere.

Second, the southern hemisphere circulation appears to be the more vigorous of the two. This is not simply a function of the difference in frictional effects. In the northern hemisphere summer, the large mass of Asia, and to some extent North America, both largely situated in middle and high latitudes, warm up and reduce the temperature contrasts between polar and tropical areas. This has the effect of weakening the power of the circumpolar vortex of the hemisphere. However, at the same time in the southern hemisphere, winter temperature contrasts are very strong and the circulation vigorous. Thus when the thermal equator migrates northwards in June and July into the northern hemisphere, the southern hemisphere circulation is able to encroach well over the equator. In the reverse season, there is no comparative warming up of large land masses in the southern hemisphere to weaken the circulation there. One noticeable result of all this is that the migration of the ITCZ into the southern hemisphere is strictly limited, whereas in the northern hemisphere summer, planetary circulation factors initiate substantial movements of the zone, enhanced in some cases by regional factors.

Topographic Barriers

Most mountain chains over three or four kilometres in height can materially affect primary circulation features. It is well known that north–south orientated topographic barriers obstruct zonal flow, and at the same time channel meridional flow. The best example occurs in the case of the Rockies, which effectively prevent a great deal of west coast marine air from penetrating inland, and conversely protect the western seaboard from cold air invasions originating in Arctic Canada. East–west aligned mountains, such as the European Alps, similarly play a role in preventing large latitudinal movements of air in the lower troposphere. The Himalayas, in particular, form an extremely effective barrier to the northward movement of the summer monsoon in India.

The effect of mountain barriers can sometimes be even more fundamental in that the highest chains materially interfere with the upper circulation. Especially important in this respect are the Rockies and Andes, which extend well up into the upper westerlies in their respective hemispheres. Air approaching these mountains on their western sides is forced to contract vertically, and therefore becomes denser and forms a ridge of relatively high pressure in the upper zonal flow. On the lee side, the reverse obtains: the air column expands, causing a trough of low pressure and encouraging convergence. The lee of the Rockies thus becomes one of the favoured locations of a Rossby wave trough. Some authorities suggest that the Rockies play a primary role in the geographical location of many of the Rossby waves, and that the wave motion, once initiated in the lee of the Rockies, inevitably produces a ridge over the Atlantic and a trough over eastern Europe. More certainly, the establishment of a high-level trough near the Rockies does encourage cold air to move southwards in the continental interior of North America, and this helps to counteract the development of any large summer monsoonal effect in the continent. By analogy, similar reasoning can be applied to the Andes and South America, where the effects of summer overheating on the shifts of the ITCZ are slight.

In Africa, on the other hand, the tabular relief is not sufficiently high to interfere with the zonal flow, and local summer heating does emphasise the migration of the ITCZ. In the case of the Himalayas, the seasonal migrations of the northern hemisphere upper air flow are substantially modified, and we can develop this point further in the consideration of the dynamics of the Asian monsoon.

Ocean Currents

Strong connections exist between the atmospheric and oceanic circulations. The rotation of winds around the major pressure systems located over the oceans, particularly the subtropical cells, is reflected in the flow of the ocean currents. These currents are capable of transporting large amounts of heat and they augment the flow of energy in both the Hadley cell circulation and also that of middle and high latitudes. An important feedback process operates here, for the currents in turn transmit their temperature characteristics into the atmospheric circulation. A consequence of this is that ocean currents can materially affect large-scale climates.

The relative warmth of north-west Europe for its latitude, and the coolness of Labrador, are both largely attributable to the effects of respective ocean currents. Other examples will be referred to in Chapter Fifteen in connection with specific climatic types. The net result is that heat transport by currents favours equability over the globe, and generally currents have an important role in helping the general circulation to ameliorate temperature contrasts between the tropics and polar areas.

The Role of the Monsoon

Finally, in this chapter, some consideration must be given to the part played by the Asiatic monsoon in the maintenance of the general circulation. Although in planetary terms it is sometimes regarded as a secondary feature, the monsoon is of very great extent and makes an excellent case study illustrating the interaction of several of the features we have been discussing in this chapter. The usual classic thesis of the origin of the monsoon mainly stresses the importance of regional factors: namely, that a summer low-pressure cell develops over land because of thermal overheating, and this induces moist air to blow inwards from ocean areas. In winter, according to long-held theories, the system is simply reversed, and cold air blows outwards from the Asian continent. However, there are a number of inadequacies in this theory; for example, it is difficult to see why a thermal low over northern India should cause the summer monsoon to penetrate as far north as Mongolia. We have also seen that thermal lows in North and South America do not result in vigorous monsoons. The monsoon is best considered nowadays in the overall context of the seasonal migration of the general circulation, with thermal and physiographic factors superimposed on it. It is also important to recognise that there are both winter and summer elements in the monsoon, since strictly speaking the word monsoon means 'wind reversal', indicating two distinct wind regimes.

Over Asia in winter, when north–south temperature contrasts are at their most prominent, westerlies dominate the upper air flow over the whole continent, including peninsular India (Fig. 13.5a). The locations of the principal jet streams appear to be substantially influenced by the great curve of the Himalayan mountain chain. A powerful jet occupies a semi-permanent position just south of the foothills, probably being related to a strong thermal gradient there, and another westerly jet skirts round the north of the Tibetan plateau. These two jets become confluent east of the Himalayas in China. The southern jet is important in steering winter depressions through northern India. Winds at this time in south-east Asia generally blow from a northern sector, developing from air subsiding beneath the upper westerlies.

In spring the westerlies begin their northwards seasonal migration, and with this, the more northerly of the jet streams intensifies at the expense of its Himalayan counterpart, although they both maintain their mean locations. The increase in solar insolation establishes thermal low pressure over north-east India, but this does not materially affect the upper flow. The low-pressure cell in fact reaches its lowest average pressure in May, although the monsoon does not break until several weeks later. This particular point strongly hints at

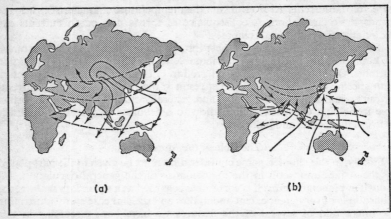

Fig. 13.5. Surface (solid lines) and upper (broken lines) airflow in the Indian monsoon region.

the importance of planetary circulation factors, rather than local topographic control, in initiating the actual outbreak. Modern evidence has now reasonably established that the summer monsoon only finally advances when the jet stream, having been prevented from systematic retreat by the Himalayan chain, rapidly disappears from its normal winter position in the space of a few days. This disappearance effectively marks the change of the upper air circulation to its summer pattern (Fig. 13.5b), which results in the establishment of equatorial upper easterlies, and the surface south-westerlies, bringing the moist monsoon. We may see this in the context of the northward migration of the equatorial trough and ITCZ, and the encroachment over the Indian subcontinent of the southern hemisphere Hadley cell, which we have referred to earlier in the chapter. This situation prevails for three or four months, until in late September and October the equatorial trough weakens in its position northernmost and retreats south. The surface south-westerlies then give way to the north-east trades, and a new ITCZ establishes itself near the equator. At high level, the Himalayan jet stream reappears rather abruptly in mid-October and the zonal westerlies return over the Indian subcontinent.

The interaction of the polar and tropical parts of the northern hemisphere circulation on the one hand, and the southern hemisphere circulation on the other, is clearly critical in all this. Some authorities have concluded that the annual pattern of the southern circulation is fundamental to the precise timing and extent of the Asian monsoon, and that the northern hemisphere in fact plays a rather passive role.

Suggested Further Reading

Barry, R. G., 'Models in meteorology and climatology', in *Models in Geography*, edited by Chorley, R. J., and Haggett, P., Methuen, London, 1967.

Hare, F. K., 'Energy exchanges and the general circulation', *Geography*, 1965, **50**, 229–241.

Pedalaborde, P., *The Monsoon*, Methuen, London, 1963.

Reihl, H., 'General atmospheric circulation of the tropics', *Science*, 1962, **135**, 13–22.

CHAPTER FOURTEEN

WEATHER

The macro-scale features of the general circulation provide the necessary framework within which we may consider the smaller scale (meso-scale) components that make up our daily weather. All areas of the world experience short-term weather disturbances from time to time, but none more so than the mid-latitude regions lying in the westerlies belt. In fact, this belt can be said to be characterised by them: here weather often changes rapidly within a matter of a few hours.

The fascination of the daily weather has stimulated a great deal of conversation, description and scientific study. The nineteenth century saw the beginnings of the growth of national weather services, and with it, attempts at forecasting and explanation. The differences between the weather of **depressions** and **anticyclones** were soon appreciated, but it was not until 1917 that the Norwegian meteorologist Bjerknes formulated a description and interpretation of the weather in terms of **air masses** and **fronts**, expressions that are familiar to most of us today. Although, with the passage of time, new information has caused some of Bjerknes' original concepts to be modified, they have not been replaced, and still form a fundamental part of the theory behind weather study. The ideas of Bjerknes have also stood the test of time as good teaching models, and form an important link between the layman and the professional meteorologist.

Air Mass Weather

The original concept of an air mass was a large body of air whose physical properties, particularly those of temperature and humidity, were uniform over considerable areas. Nowadays this has been modified slightly, and we think of air masses as areas in the atmosphere where horizontal gradients of the main physical properties are fairly slack. The term can be properly applied only to the lower layers of the atmosphere, but the horizontal extent of an air mass may extend to several hundred thousand square kilometres.

As noted in Chapter Ten, the atmosphere tends to acquire many of its properties via the Earth's surface, and hence air masses derive their temperature and humidity levels mainly from the region over which they lie. These areas are known as **air mass source regions**. Some regions of the globe are more likely to give rise to characteristic air-mass types than others. Two conditions govern the occurrence of the principal source regions. First, they occur in regions of relative calm in the general circulation, notably the semi-permanent anticyclones, where air motion is sufficiently slow for homogeneous air-mass characteristics to develop. Second, source regions are usually areas where the Earth's surface itself is fairly uniform, such as oceans, or deserts, or large ice- and snow-covered areas. Although air masses can become considerably modified as they move away from their source regions, it is usual to describe the main types of air mass in terms of their region of origin.

Principal Air-mass Types

The most widely used classification of air masses uses a simple symbolic nomenclature, as shown in Fig. 14.1. There are two major groups: **Polar** air masses, represented by capital **P**, and **Tropical** air masses, denoted by **T**. The Polar Front represents the fluctuating boundary zone between these two types in middle latitudes. We may note also that the 'polar' air masses actually originate in cool temperature or sub-arctic regions rather than the polar areas themselves. These two main groups include nearly all air masses likely to be encountered in middle latitudes. However, additional groups are sometimes

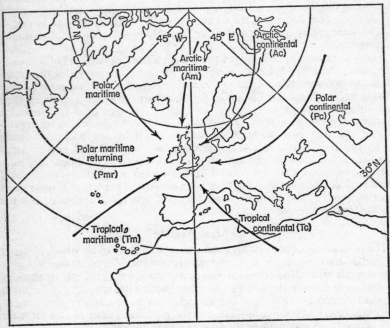

Fig. 14.1. Air masses affecting the British Isles.

designated, notably **Arctic** (A) and **Equatorial** (E) air masses, which occasionally migrate well beyond their original source regions.

Each of these major groups may be further subdivided according to whether the source region is **maritime** (m) or **continental** (c) in character. This distinction reflects something of the considerable humidity differences that may occur in air masses, depending on whether they originated over oceans and acquired typical maritime properties of mild temperatures and considerable moisture content in their lower layers, or over continents, where they are likely to be relatively dry.

In the northern hemisphere, **Polar continental** (Pc) air masses have their source regions over central Canada and Siberia. In winter these are some of the coldest places on Earth, and air masses emanating from here are extremely cold, very dry, and normally cloud free. The migration of these air masses can bring some of the coldest weather experienced in otherwise fairly equable areas, such as the British Isles. In summer, a certain amount of warming of the lower layers of this type of air mass makes them less stable and more cloudy. **Polar maritime** (Pm) air masses originate at high latitudes over the northern Pacific and Atlantic oceans, and are essentially cool and moist and relatively unstable in their lower layers. These air masses are frequent visitors to north-west Europe and the British Isles, bringing dull rainy conditions.

At much lower latitudes, the subtropical high-pressure centres over oceans act as major source regions for **Tropical maritime** (Tm) air masses. These are typically warm, moist and unstable, especially in summer when convective instability is frequent. **Tropical continental** (Tc) air masses originate over warm desert areas, such as the Sahara and the arid parts of the United States. Not unexpectedly, these air masses are hot, dry and unstable, especially in summer, although they are too deficient in moisture to cause cloud development.

Modification of Air Masses

Air masses move away from their source areas in accordance with the pattern of the general circulation. During migration, they may have their essential characteristics substantially modified, so that the weather they eventually bring may be rather different from that in their source areas.

Changes may take place in two principal ways: either by internal modification—for example, by subsidence, bringing about important adiabatic changes (Chapter Eleven)—or through the effect of the surface areas over which they pass. The end-result of either process is to produce **secondary air masses**, which in terms of the weather of the British Isles are probably more important than unmodified air masses. Migrating air masses which are warmer than the surface over which they are travelling are given the suffix **w**; thus Tmw represents a warm Tropical maritime air mass which is gradually becoming cooler and therefore more stable in its lower layers as it passes over colder areas. Air masses modified in this way are often characterised by stratiform cloud or fog. On the other hand, air masses which were originally cooler than the surface over which they are passing become increasingly unstable as their surface layers are warmed. These air masses are given the suffix **k** (*kalt* = cold). Generally speaking, cool air masses are usually more subject to modification than warm air masses.

One typical example of an air-mass type which undergoes substantial modification is Pc air originating from the high-latitude interior of North America. This air frequently travels westward across the Atlantic towards Europe and is considerably modified by the Gulf Stream, leading to an increase in temperature and moisture in its lower layers, and transforming it into a Pm-type air mass. Similarly, eastwards incursions of Pc from Eurasia towards the British Isles are modified by their passage across the North Sea, but in this case the degree of modification is much less, and the relative warmth of the sea has only a marginal effect on the inherent cold temperatures of the air mass.

The overall concept of air masses works very well for practical purposes of explanation in mid-latitude zones, but it is more difficult to apply in equatorial regions, where air-mass contrasts are less well marked. The main distinguishing characteristics that do develop in low latitudes are better related to features of vertical air motion in adjacent subtropical cells, rather than to mass horizontal migrations of air.

Fronts

Fronts are broad **mixing zones** in the atmosphere where horizontal gradients of pressure, temperature and other properties of air masses become steepened. On a weather map, we conventionally represent a front as a line; this is partly because a front was originally regarded as a relatively narrow zone separating two air masses. The conventional symbols used for fronts can be found in Fig. 14.5.

Although not all fronts are the result of simple air-mass convergence, in their formation they do tend to be restricted to areas of strong pressure gradient, often brought about by thermal differences. Thus favoured locations for the development of fronts include the margins of oceans and continents and the snow and ice margins of high latitudes. Generalised positions for some of the world's major fronts are given in Fig. 13.1. As one would perhaps expect, the most closely studied fronts occur in the Polar Front system of the northern hemisphere, which affects much of North America and northern Europe, including the British Isles. Most of the discussion below concerns these fronts, but very similar systems also occur in the westerlies belt of the southern hemisphere.

Frontal Structure

The depiction of a front as a line on a map also belies the fact that frontal zones have a considerable vertical extent, with discontinuities reaching right up to the tropopause. In cross-section (see Figs. 14.2 and 14.3) the frontal zone slopes upwards at a low angle, with the colder of the two air masses forming a wedge underlying the warmer air mass. We generally exaggerate the slope of the frontal surface in a diagram, but the slope is actually very small, often less than 1°. Thus in a situation where a frontal zone stretches some 8–9 km up into the troposphere, it may have a horizontal coverage of the ground of up to 1000 km. Hence, an observer may well see the symptoms of an approaching front many hours before the surface front actually reaches him.

In the frontal zone, not only are there liable to be temperature changes, but wind direction and speed also change. Wind velocity often increases with height, and when temperature differences are particularly great, jet streams may form, often of great velocities. The relationship between the jet stream and frontal patterns is indicated in Fig. 14.4. Such a jet stream associated with a front can sometimes be recognised by long parallel **streets** of high clouds.

Types of Front

Fronts are rarely stationary, but move with the general circulation, bringing rapid sequences of weather change. The weather associated with them is extremely variable; the common denominator is that it is generally unsettled. Two factors largely determine the weather experienced: namely whether the front is warm or cold, and the degree of activity on the front. This twofold criterion gives us a basis for describing types of front. Where a front passes us in which cold air is replaced by warmer air, we have a **warm front** situation; where cold air comes second, we have a **cold front**.

The degree of activity at both types of front is determined largely by the vertical motion of air in the warm air mass. If the air here is unstable and rising rapidly, the front is usually very active, and these are termed **ana-fronts** by meteorologists. On the other hand, the general sinking of warm air at the front suppresses weather activity, and these are called **kata-fronts**. Although individual frontal situations can vary very widely, the following descriptions can be regarded as typical models.

Warm Front Weather

The ana-type of warm front is by far the most common warm-front type in Britain. The relative air motion at the front, indicated by arrows (Fig. 14.2a), is important in determining the weather. The rising motion in the warm air mass often proceeds at different rates at different levels, and this, together with varying relative humidity, produces a multi-layered effect rather than one solid mass of cloud. The clouds at higher levels do not always coincide with the frontal zone as defined by changes in temperature. The relative humidities in the cold air mass, especially near the frontal zone, are often very low. Here air motion is predominantly downwards.

The whole front will be moving at an average speed of about 50 km/h (30 m.p.h.). The weather sequence we can expect to experience can be deduced by

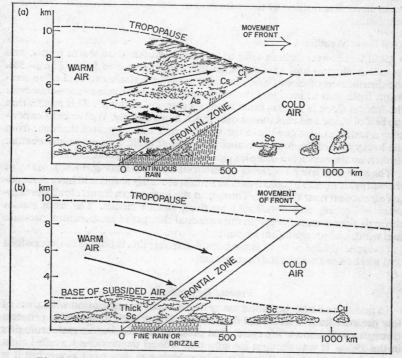

Fig. 14.2. Cross-sections through typical warm fronts; (a) ana-type;
(b) kata-type.

moving the diagram to the right. The normal sequence of clouds is indicated: cumulonimbus and cumulus cloud in the cold air mass (Pm) well ahead of the front, dying out as it approaches; isolated cirrus in the warm air above the front, gradually increasing and merging into cirrostratus; then a thickening and lowering of these clouds to altostratus, obscuring the sun and giving precipitation. Near the surface front, nimbostratus predominates, giving persistent rather than heavy rain. **Frontal fog** may form in the cold air mass near the frontal zone as this moisture condenses. The precipitation behind the frontal zone in the warm air mass usually comes from stratocumulus clouds in the form of drizzle, gradually dying out and being replaced by unsettled weather typical of the warm air mass.

As the front passes through, the change of air mass causes a temperature rise, which may be of the order of 4–5°C, over the period of one or two hours. Pressure levels, which would have been falling steadily as the front approached, recover slowly, and the wind usually veers (clockwise) as the front passes through.

On **kata-warm fronts** (Fig. 14.2b), the downward motion of the air near the front in the warm air mass considerably restricts the development of medium and high-level clouds. Hence thick stratocumulus clouds prevail, giving light rain or none at all. Although changes in temperature, pressure and winds occur in similar fashion to those for an ana-front, they are usually smaller in amplitude.

Cold Front Weather

Cold fronts over Britain tend to be more variable than warm fronts, and include both ana- and kata-types. In the 'classic' **ana-cold front** (Fig. 14.3a), the distribution of clouds and precipitation is very similar to that of an ana-warm front, but in reverse. One other important difference occurs, however: the frontal slope is two or three times greater, i.e. about 2°. This means that the belt of cloud and rain passes over much more quickly. We typically experience a rather abrupt change in the weather as the front passes through, from the heavy rain of the nimbus clouds at the front itself, to clear bright weather broken by showers from cumulus clouds in the following cold air mass.

The **kata-cold front** has rather different characteristics (Fig. 14.3b). Many of the warm air masses at cold fronts in Britain are fairly stable and have descending air motion near the front. Thus, as in the kata-warm front, cloud development is limited, and dominated by thick stratus clouds. The front passes through almost unnoticed with only gradual changes in temperature, pressure and wind, and precipitation amounts.

One special kind of front is the occluded front; this is unique to depressions and will be dealt with in the next section.

Frontal Depressions

In middle and high latitudes, many fronts occur in association with areas of low pressure known as **frontal depressions**. These, with their fronts, progress through a well-known 'life cycle', first proposed by Norwegian meteorologists in the 1920s. It was thought initially that frontal depressions travelled right round the world on the Polar Front, developing as they went along, but it is now recognised that the life of a system is typically 4–5 days.

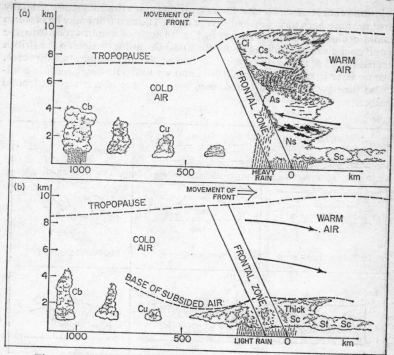

Fig. 14.3. Cross-sections through typical cold fronts: (a) ana-type;
(b) kata-type.

Development of a Frontal Depression

The sequence of events depicted in Fig. 14.4 starts with the existence of two contrasting air masses on either side of the Polar Front, with cold air to the north and warm air to the south. The first sign of a developing depression occurs with the formation of a **wave distortion** on the front (a). The apex of the wave becomes the centre of the pressure area, and with the increasing amplitude of the wave, pressure falls rapidly and closed isobars may be drawn. The development of the wave traps warm air in the **warm sector** of the depression between the cold air in front and behind.

At this stage of the depression (b), with the system moving eastwards in accordance with the general circulation, on the ground we experience a typical weather sequence: the leading edge of the wave creates a warm-front weather pattern similar to that previously described, followed after some hours by cold-front weather on the trailing edge.

The majority of lows which reach the warm sector stage continue to deepen and become more vigorous. The cold front travels faster than the warm, overtaking it first of all near the centre of the low and then progressively further outwards. This process is known as **occlusion**, and is clearly associated with the raising aloft of the warm air (c). A new front is nearly always formed in the occlusion process, since by now the two air masses in the cold sector, initially the same, have been modified in different ways. The new front may be of warm or cold occluded type (that shown in Fig. 14.4c is of a cold type, with the second cold air mass wedging under the first). Occlusions reaching the British Isles tend to be of the warm type in winter, and cold in summer. In either case, since we have in effect the original cold and warm fronts raised aloft in addition to the new surface occlusion, they give us some of our wettest frontal weather.

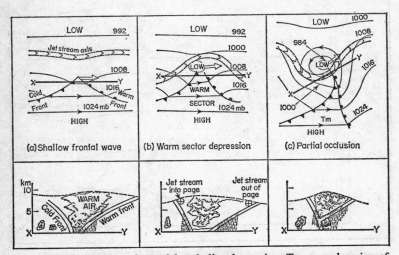

Fig. 14.4. Development of a model occluding depression. *Top row:* plan view of jet stream axis at 300 mb superimposed on surface pattern. *Bottom row:* cross-sections along lines shown in top row.

The system usually begins to decay when the occlusion is complete; the occlusion itself frequently becomes detached from the centre of the low, which turns off to the left of its original track and slows down. The low may fill up rapidly if a more vigorous depression moves into the area. Some depressions, on the other hand, persist unchanged for days.

Depression Patterns

Frontal depressions rarely occur in isolation, and surface weather charts frequently reveal complex patterns of frontal waving. It is common for a whole string of frontal depressions, often at different stages of development, to occur on the Polar Front. Such a series, usually up to seven in number, is known as a **depression family**. One such example in the North Atlantic is presented in Fig. 14.5. These can give spells of unsettled weather lasting a week or more. The sequence is eventually terminated because with each successive depression, the Polar Front is pushed further and further south of its usual position, and eventually cold polar air forms an extensive wedge of high pressure.

Secondary depressions commonly occur within the circulation of the main depression. They may form at the point of occlusion; other secondary depres-

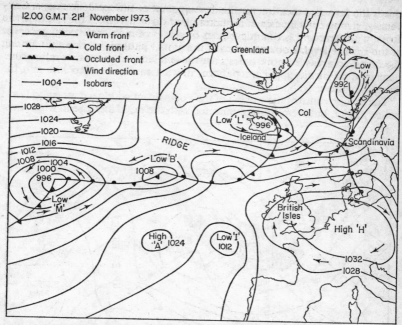

Fig. 14.5. Depression family in the North Atlantic, November 21, 1973. Note different stages of development: Low M, a warm sector depression; Low L, partially occluded; Low K, a fully occluded type.

sions develop as the result of renewed waving on the trailing cold front. In many cases these move rapidly along the front and are absorbed by the parent depression, but they may have the effect of causing a local intensification of precipitation and often delay the clearance of the front itself. In other cases, secondary depressions may mark the beginning of a new main depression.

The Origin of Frontal Depressions

Although the stages in the development of a frontal depression are fairly well known, its initiation (**cyclogenesis**) has been something of a puzzle in the past. New evidence has come forward in recent years with the closer examination of the upper air situation.

We observed in Chapter Twelve that surface depressions can only develop where divergence at high levels in the atmosphere removes air brought into the depression at lower levels (Fig. 12.6). In the Rossby wave pattern in the upper air, acceleration and deceleration of air is continually taking place as the waves meander first equatorwards and then polewards. Thus the polar jet stream is vigorous in some places and not in others. In conjunction with these speed changes, there is compensatory **cross-isobaric flow**, causing convergence in some places and divergence in others (Fig. 14.6). This is sufficient to generate and maintain cyclones and anticyclones at surface level underneath. Divergence tends to occur downstream of a trough in the upper Rossby waves, and this is the favoured location for the development of surface depressions. Conversely, strong convergence of air takes place downstream of the ridges in the Rossby pattern, and here anticyclones or ridges of high pressure may persist at the surface level. Looking again at the global pattern, we find that among the favoured areas for cyclogenesis are those related to the main semi-permanent troughs in the upper air pattern, downstream of the Rockies and the Tibetan plateau.

The precise cause of the waving of the Polar Front within a depression is not entirely clear, but seems to be connected with the rapid increase in wind shear

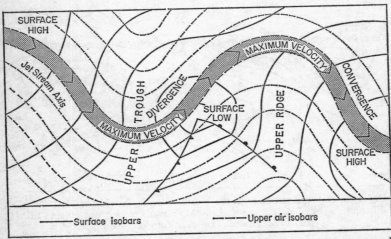

Fig. 14.6. Surface pressure systems and their relation to convergence and divergence in the upper zonal flow. This figure should be compared with Fig. 14.4.

with height. This takes a spiral-like form and causes the front to twist itself into wave patterns.

Frontal depression systems follow the flow of the upper air eastwards in middle latitudes but at a lower speed, amounting to about 70 per cent of the geostrophic speed in the warm sector. Thus analysis of the upper air information helps us to forecast the weather chart for the following day.

Non-Frontal Depressions

By no means all depressions originate as frontal waves. There are several other possibilities to explain why surface low-pressure systems should develop. Whatever their cause, all depressions share in common the characteristic of ascending air, which in many cases inevitably leads to cloud and precipitation.

Thermal Depressions

Shallow lows often develop because of local overheating of part of the Earth's surface. The scale of this mechanism can vary greatly. Small diurnal examples may occur over islands and peninsulas in summer or over lakes and seas in winter. At the other extreme, continental low-pressure cells develop in summer over the large arid regions of the world, notably in central Australia and the south-west United States. In Britain and similar mid-latitude areas, small thermal lows generally occur over land in summer when insolation is highest. The weather associated with these depends largely on the air mass in which they form. In a Tc or Tm air mass, the weather may be hot and dry; but in Pm conditions, showers and thunderstorms are more likely.

In winter, a different variety of thermal low occurs in polar air, largely because of heating over the sea. These are sometimes known as polar lows, or in weaker cases they form polar troughs. These lows are usually of greater intensity than the summer thermal low, and may give prolonged rain or snow.

Lee Depressions

These are sometimes alternatively known as **orographic depressions**. They form when the flow of an air stream is affected by a mountain barrier, causing a piling up and divergence of air on the windward side, and a relative deficit and convergence on the lee side. This deficit is often sufficient to show itself on a chart as a shallow low or trough.

British mountains are not really high enough to create good examples of lee depressions, but they are common in Europe, especially in north Italy, south of the Alps, and they are also found to the east of the Rockies. This type of low is essentially non-frontal in origin, but near-by fronts may be drawn into the circulation.

Tropical Depressions

In low-latitude parts of the world, intense depressions with a very low central pressure play an important role in the weather. They are variously known as hurricanes (Caribbean), cyclones (Indian ocean), typhoons (China Sea), and willy-willies (Australia). An occasional hurricane may affect Britain.

A simplified idea of the structure of a **hurricane** (Fig. 14.7) shows that right at its centre there is a warm central core or eye, largely cloudless, warmed by descending adiabatic winds which are being sucked downwards to supply the surrounding vortex. Despite the very low pressure in the eye, it is surprisingly calm. Around the eye is a great cylinder of clouds, torrential rain and violent

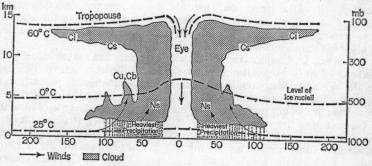

Fig. 14.7. The structure of a hurricane.

winds. Here great amounts of latent heat are released which act as the main energy source to drive the system. The strongest winds are found 15–20 km from the centre; beyond this they become progressively less.

The precise trigger mechanism which sets off such a system is uncertain: some hurricanes may grow from pre-existing disturbances; but they seem to form only under certain conditions. They originate over sea areas where the surface temperature is greater than 27°, i.e. largely in the tropics; but they rarely develop within 5° of the equator where the Earth's deflecting force is too small for a circulation to develop. There also seems to be some relationship between their formation and the position of the equatorial trough: for instance, in the Caribbean the hurricane season is in August and September when the equatorial trough is in this region. The same applies to the western Pacific near Australia.

Hurricanes move slowly, westwards at first in compliance with the general circulation, but then they curve polewards, taking on even greater cyclonic rotation, and eventually eastwards on reaching middle latitudes. They usually die out on reaching cooler waters or land, but before doing so, some hurricanes cause great damage to coastal areas. Occasionally hurricanes may be revitalised as mid-latitude circulations and in this way eventually reach areas such as the British Isles in modified form.

The ultimate stage of concentration of cyclonic spin and energy occurs in **tornadoes.** These are much smaller than hurricanes and, unlike them, form over land. Overhead, there is typically a large cumulonimbus cloud, with a **funnel cloud** projecting from its base to the ground. The great pressure fall in the whirl causes cooling and saturation and this makes the funnel cloud visible. It is thought that the tremendous rotation in tornadoes may be initiated by the rapid convergence of air at the base of the cumulonimbus cloud, drawn in as rapid updraughts develop.

Tornadoes are common over the Great Plains of the United States. Perhaps surprisingly, they also affect Britain about one day per month on average. They are not normally severe. One recent example which caused considerable damage occurred on September 26, 1971, in the Rotherham area.

Anticyclones

An anticyclone is an area of high pressure, with roughly concentric isobars and clockwise circulation of winds, and, as the name suggests, is the meteorological opposite of the depression. There are also a number of other differences: the degree of high pressure in an anticyclone rarely matches the intensity of low pressure in a depression; anticyclones usually cover a wider area; and they also tend to be more persistent and slow moving.

It was noted in Chapter Twelve that anticyclones are maintained by the subsidence of air throughout much of the depth of the troposphere. This provides the key to weather conditions in an anticyclone; stability prevails, with only light winds or even calm at the centre of the system. However, anticyclonic weather is not uniformly fine; much depends on the type of air mass and the season of the year. In winter, some anticyclones give clear skies and low temperatures, and severe frosts may occur. Others, when moist Pm air is present, may form continuous stratus cloud, trapped beneath an inversion formed in the subsiding air above. In the recent past, such spells of **anticyclonic gloom** have sometimes combined with the products of industrial and domestic combustion to create considerable pollution problems in certain large cities in Britain and elsewhere.

By way of contrast, most anticyclonic spells in summer in mid-latitude areas are usually fine, and often give us our maximum temperatures of the year.

The Origin of Anticyclones

The high pressure in an anticyclone is a direct result of the presence of cold dense air somewhere in the system. On the basis of where this cold air occurs in the vertical structure of the anticyclone, they are sometimes classified into warm and cold types.

In **cold anticyclones**, the cold air is confined to the lowest parts of the atmosphere—say within 2 km of the ground. A simple thermal mechanism accounts for the cold air: it is caused by the cooling of the Earth's surface in winter, and this in turn chills the adjacent air, causing it to contract and initiate the subsidence above. This mechanism is well seen in the great seasonal anticyclones which develop over Siberia and Canada in winter.

On the other hand, **warm anticyclones** are characterised by relatively warm air in the lower parts of the troposphere. The excess of pressure seems to arise from the coldness of air in the upper troposphere and lower stratosphere. The permanent subtropical anticyclones are of this type. Some of the warm variety of anticyclones which affect Britain originate as tongues, or pools, of warm subtropical air which encroaches northwards in a low-index upper air situation (see Fig. 13.4).

Not all anticyclones readily fall into these two types. The highest mean sea-level pressures of all, over 1070 mb, occur in continental winter anticyclones when cold air at low levels combines with cold air in the higher troposphere. In another case, some anticyclones over Britain gradually change from cold to warm type as the lower layers of the atmosphere warm up.

Blocking Situations

Anticyclones in the westerlies belt, whether they are of cold or warm type, vary considerably in their persistence and scale. Some are little more than ridges of high pressure between successive depressions. Others may last several days or even weeks.

The longer lasting anticyclones occur when the upper air flow adopts a low zonal index, breaking down into a cellular pattern. Under these conditions blocking anticyclones may develop. This means that the surface anticyclone, influenced by the upper air pattern, blocks the passage of surface depressions and steers them well to the north or south. Blocking anticyclones may occur at any time of the year and often produce marked weather anomalies.

A notable example which affected the British Isles occurred during the winter of 1962–63, when a blocking anticyclone was centred south-east of Iceland for several weeks in January and February. Persistent north and north-easterly winds blew over Britain, bringing heavy snowfall and some of the lowest mean daily temperatures recorded since 1740. A more recent, and perhaps more typical example is depicted in Fig. 14.8, for March 1973, again under the influence of a low-index situation.

Weather Prediction

The task of predicting the weather relies basically on the recording of atmospheric conditions over large areas at close intervals of time. This information is then used in various ways to predict what is likely to happen next. Traditional methods of forecasting use the data to produce **weather maps** of various kinds for both surface and upper air conditions. Simplified versions of the daily weather chart appear regularly in many newspapers and on television. With the aid of both maps and data, the weather-forecaster tries to estimate the speed and direction in which individual depressions, anticyclones, air masses and fronts are likely to move, and what weather they are liable to bring with them. But a great deal of experience and intuition is needed on the part of the forecaster to estimate the time factor involved: the rates at which these weather phenomena grow and decline vary considerably from day to day. Thus, in practice, despite great advances in recent years in information gathering through the use of satellites and radar, weather prediction using

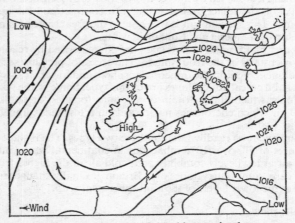

Fig. 14.8. Surface chart of blocking anticyclone over Britain, March 15, 1973. This persisted for almost three weeks, giving warm, rather cloudy weather.

weather maps is rarely possible beyond 48 hours. Within that time, however, good results are generally obtained.

New methods have been evolved in recent years to support traditional forecasting procedures. In particular, **mathematical methods** use basic physical laws to compute the future condition of the atmosphere on the basis of its present state. High-speed computers are necessary to cope with all the necessary calculations. Results employing these methods have proved successful for short-range forecasting. For long-range weather prediction of a week or more, two principal methods are in current use. One of these attempts to extrapolate **statistically** the likely behaviour of events in the upper atmosphere at the critical 700 mb level. Thirty-day forecasts based on this method have been used in the United States since 1948. The other method, more favoured in Britain, uses **analogue models**, based on the principle that sequences of weather are likely to follow a similar course to one in the past if initial conditions are similar. The problem is then to search the records for a situation closely analogous to the one prevailing.

Despite the wealth of information available about the weather at any one time, it has been well said that weather prediction is not an exact science, and the complexity of the atmosphere makes it unlikely that it ever will be.

Suggested Further Reading

Barrett, E. C., *Viewing Weather from Space*, Longmans, London, 1967.

Pedgley, D. E., *A Course in Elementary Meteorology* (Chapters 9–12), HMSO, London, 1962.

Riehl, H., 'On the origin and possible modification of hurricanes', *Science*, 1962, **135**, 1001–1070.

Sawyer, J. S., 'Weather forecasting and its future', *Weather*, 1967, **22**, 350–359.

Taylor, J. A., and Yates, R. A., *British Weather in Maps*, Macmillan, London, 1958.

Wickham, P. G., *The Practice of Weather Forecasting*, HMSO, London, 1970.

CHAPTER FIFTEEN

CLIMATIC TYPES

Climate has sometimes been defined as 'the average state of the atmosphere', but such a definition is really only a partial one. It gives a rather static impression and tends to disguise the constant variations of the weather. A better definition might be to think of climate as 'an association of weather variations'. If we describe the climate of any particular place we need to consider not only the statistical record, but also the dynamic background to the figures.

As part of the basic information needed when seeking to recognise climatic types, the global distribution of the main climatic **elements**—namely, temperature, precipitation and winds—must be taken into account. We can recall here some of the earlier charts in this book of mean January and July temperatures (Figs. 10.6 and 10.7), the distribution of precipitation (Fig. 11.9), and pressure patterns and planetary winds (Fig. 12.1). It is also worth re-emphasising that several **factors** have a fundamental controlling influence on the distribution of these elements, notably the radiation budget, the inclination of the Earth, and the general circulation. Important **geographical factors**, which will be illustrated in several examples later in this chapter, include latitude, altitude and the distribution of land and sea. If we were to consider local climate, more factors such as the inclination and aspect of slopes would come into play, but these will be looked at in Chapter Sixteen.

Climatic Classification

Climatic classification rests on the idea that, despite the seemingly infinite variety of climates, patterns of weather and elements do repeat themselves in various parts of the globe where the essential governing factors are similar. For example, west-facing coastal areas between 40° and 60° of latitude all tend to experience a similar mild oceanic climate (north-west Europe, British Columbia, southern Chile, Tasmania and New Zealand). Many attempts have been made to classify climate, some of them going back into ancient history. The Greeks, for instance, spoke of the world as being divided into three zones: torrid, temperate and frigid. More recent classifications have largely concentrated on defining meaningful boundaries between climatic zones.

Most classifications naturally make use of the climatic elements in some way as a basis for categorisation. Classifications simply based on one single element would clearly be rather misleading: the Greeks saw a basic parallelism between temperature zones and latitude, but humid and desert regions received no distinction. Equally, if precipitation were used by itself, a climatic category defined by the characteristic of low precipitation would include both hot deserts and arid polar areas. If we were to combine the various elements, this may well produce a more satisfactory classification but what values or boundaries should we choose? We can divide the various approaches that have been attempted to this question into two broad categories. First, classifications whose zones are based on the *effects* of the climatic elements, and second, classifications whose zones are defined according to the *causes* of the climate.

Classifications based on Effects

Several classifications have been proposed on the idea that natural vegetation zones should be a faithful reflection of climate. Thus vegetation boundaries have been taken as the lines at which to fix climatic boundaries, expressed in terms of temperature and rainfall amounts. However, it has not proved a simple task to recognise natural boundaries, still less to ascribe climatic data to them. Not only are most vegetation boundaries actually very gradual transitions, but the effects of both human interference and local rock and soil conditions make it difficult to assess the status of the vegetation boundary in relation to climate. Moreover, continuing climatic change (Chapter Seventeen) makes it likely that some vegetation zones are not entirely a reflection of the prevailing climate.

Despite these difficulties, some of the most widely used classifications are of the 'effect' type, such as those employed by W. Koppen and A. A. Miller. These achieve comprehensive coverage of the globe, but the number of categories recognised is high, in some cases necessitating the use of rather complex notation systems.

Other classifications in this broad group have attempted to relate climate to its effect on the **water budget**. Instead of simple temperature and rainfall figures to define boundaries, they use evaporation indices derived from careful observation and experiment. Unfortunately, evaporation data are not uniformly available throughout the world, and such classifications, although of considerable practical value, are not widely used.

Classifications based on Causes

If we want to understand the geographical distribution of the world's climates, it helps to understand their origins. Thus attempts have been made to construct explanatory or genetic classifications of climate. The difficulty of this approach is to ensure that the causes are fully understood. Inevitably genetic classifications use less precisely defined boundaries than 'effect' classifications, but this is perhaps a much more realistic approximation to the real world.

As more is learnt of the workings of weather and climate, new attempts have been made in recent years to employ genetic classifications or descriptions. Some authorities have defined climatic types in terms of characteristic airmass types. Perhaps the most fruitful approach is simply to use the major features of the general circulation as a basis for division, with precipitation or temperature in a secondary role. This is the approach that will be followed here; it has the advantage of making use of much of the explanatory material we have been looking at in previous chapters.

The list of climatic types in Table 15.1 is based on the categories suggested by the German meteorologist, H. Flohn. These may be compared with the world map (Fig. 15.1), which is based on a number of sources and indicates the approximate distribution of these climatic types. Some basic information is also given for a representative climatic station in each of the zones (Table 15.2).

Table 15.1.   Climatic Zones and the Atmospheric Circulation

	Approximate latitude	Circulation features
1. Equatorial rain zone	5°S–10°N	Dominated by equatorial trough. Weak variable winds mostly from west.
2. Tropical summer rain zone	10–20°N, 5–20°S	Fine and dry trade wind winters. In the summer cloudy with rain from ITCZ.
3. Subtropical dry zone	20–30°N, 20–30°S	Hot all year with weak and dry trade winds. Absent from eastern sides of continents.
4. Subtropical winter rain zone	30–40°N, 30–35°S	Fine and dry summers; subtropical high pressure. Westerlies in winter.
5. Temperate zone	40–60°N, 35–55°S	Prevailing winds from west; moderate precipitation all year. Varying continental effects.
6. Subpolar zone	60–80°N, 55–70°S	Both westerlies and polar easterlies. Heavy precipitation near oceans.
7. High polar zone	80–90°N, 70–90°S	Ice caps; variable winds, predominantly east. Little precipitation.

Table 15.2.   Climatic Data from Representative Stations for the Seven Climatic Zones

	Mean temperature (°C)		Precipitation	
Zone and location	Warmest month	Coldest month	Total (mm)	Frequency (days)
1. Singapore 1°N, 104°E	28·0	26·1	2280	175
2. Darwin 12°S, 131°E	29·2	25·1	1560	98
3. Alice Springs 23°S, 134°E	28·2	11·9	250	32
4. Lisbon 39°N, 9°W	22·5	10·8	710	116
5. Valentia (Eire), 52°N, 10°W	15·4	6·8	1400	252
6. Reykjavik 64°N, 22°W	11·2	−0·4	800	191
7. Cape Chelyuskin 78°N, 104°E	1·5	−28·0	240	No data

1. Equatorial Zone

Equatorial climates are frequently described as being uniformly hot and moist, dominated by a daily pattern of weather change which is much more regular than any seasonal rhythm. Sweaty heat, with temperatures around 30°C and relative humidities always above 60 per cent, is the general order of the day. Only at night do temperatures drop; night has sometimes been labelled 'the winter of the tropics', although the temperature drop is usually only slight. Nocturnal cooling may be sufficient to create early morning mists, but these clear quickly as the sun rises abruptly after 6 a.m. Towards midday, the increasing heat of the day triggers off convection, creating towering cumulonimbus clouds and giving heavy thundery rain in the later part of the day.

This general description is perhaps of more limited application than the impression given in some accounts. Strong diurnal rhythms certainly do exist in equatorial regions, but they are not all like the one described. Patterns of this sort are best developed in coastal areas, where local sea-breeze circulations are regularly set in operation. Many regions, such as the Malay peninsula, have very irregular diurnal rhythms, because of the variable contributions of local low-pressure disturbances, land and sea breezes, and topographic effects. Over sea areas, which make up the great majority of the equatorial zone, relatively little rainfall is received. The greatest precipitation amounts are the result of orographic effects, notably witnessed by some of the mountainous areas of Indonesia.

As a common basis, equatorial climates lie under the influence of the equatorial trough for most of the year, although with seasonal migration its dominance may vary. Coastal West Africa, for instance, has a definite rainier season, although in that it does not come under trade wind influence to any marked degree, it is best considered as having an equatorial climate. The same applies to Indonesia and surrounding countries, where the large changes in position of the ITCZ in this region allow two distinctly wetter periods to be recognised, one in June and the other in September.

But not all areas near the equator have a humid 'equatorial' type of climate of the type just described. In South America, this kind of regime is largely confined to the Amazon Basin. West of the Andes, dry trade-wind circulation from the southern hemisphere penetrates northwards to the equator across a cool ocean current, restricting an equatorial type of climate to a small area of Colombia around 5°N. On the African continent, the West African coast and much of the Congo Basin have a humid equatorial climate, but in East Africa, the high elevation of much of the terrain considerably reduces mean temperature and rainfall.

2. Tropical Summer Rain Zone

This climatic type embraces some contrasting regions of the tropical world (see Fig. 15.1), but they can all be regarded as a distinct climatic type on the basis that the weather of the equatorial trough prevails in summer, whereas in winter trade-wind conditions are in evidence. This twofold system brings marked seasons to some areas, but less so to others where the winter trades are moist.

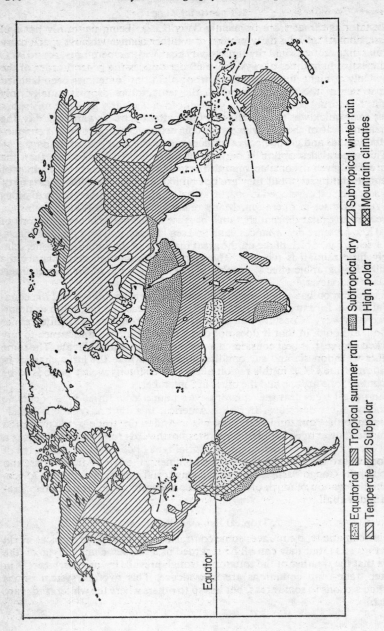

Fig. 15.1. World climatic types, based on the main circulation features of the globe.

Equator

Equatorial   Tropical summer rain   Subtropical dry   Subtropical winter rain

Temperate   Subpolar   High polar   Mountain climates

Summer Conditions

In the summer in either hemisphere (Fig. 15.2), the poleward migration of the equatorial trough brings weather conditions similar to those of the equatorial rainy zone. The rainy season usually begins and ends with heavy thunderstorms associated with the unstable conditions of the passing intertropical convergence zone. Even in areas outside the Indian subcontinent, the arrival of the rains is fairly dramatic and the term 'monsoon' can be loosely applied.

In Africa, the south-west monsoon brings welcome precipitation in June to a large tract of country stretching from the savanna lands of West Africa eastwards to Ethiopia. Precipitation reaches a maximum on the western edge of the Ethiopian mountains. To the lee of these, in the Red Sea area, much less rain falls. Similarly in the southern part of Africa, summer rain associated with the equatorial trough reaches the veldt country in December and January.

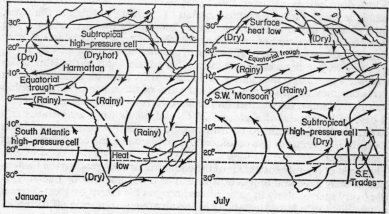

Fig. 15.2. Summer and winter conditions in Africa.

In many areas, the summer rainy period is frequently punctuated by disturbances developing along the edge of the convergence zone and travelling westwards. In the Caribbean and China Sea, these perturbations usually occur towards the end of summer; some of them grow into hurricanes.

Winter Conditions

For large areas of the savanna lands of Africa, interior Brazil and northern Australia, the summer rain provides refreshing contrast to the dry conditions of winter. In this season, trade winds blow from an easterly quarter, often having crossed hot deserts, and inevitably bring desiccating conditions. Northern Nigeria is typical: skies are generally cloudless at night, but the day becomes hot and hazy as the wind freshens. Sometimes the wind blows with considerable force; in Nigeria this is known as the **Harmattan**, and although it may relieve the humid atmosphere of the coast, it is most unwelcome in the north of the country. By the end of the dry season, northern Nigeria and many similar tropical continental areas are a scene of withered grass, parched soils and dried-up river courses.

On the eastern sides of Africa and South America, the trade winds in winter are blowing onshore, having crossed large tracts of warm water. Thus in such places as Madagascar, Natal, northern Queensland, east Brazil, and the Caribbean Islands and coasts, the winter season is not one of drought, but rather of moister conditions. Over the open sea, the trade winds generally give fine weather, with blue sky flecked with cumulus cloud. On reaching coasts or islands, orographic uplift creates rain clouds, and on the windward side of some islands, very heavy precipitation falls.

The same general circumstances are repeated on a larger scale on continental coasts; thus in effect there is no dry season to these areas. The seasonal differences are largely a matter of temperature and wind direction and force. The natural vegetation, not unexpectedly, differs little from equatorial forest.

South Asian Monsoon Region

Some descriptions consider this to be a separate climatic type, but in dynamic terms it is an exaggerated version of the tropical summer rain type, having marked seasonal contrasts. An explanation of the mechanism of the monsoon and its relationship with the migration of the equatorial trough may be found in Chapter Thirteen. Here we can consider the climatic effects.

In winter, when the westerly jet stream lies to the south of the Himalayan range (see Fig. 13.5a), the climate of India is at its most pleasant. Although depressions may affect the northernmost parts of the subcontinent, bringing winter rain or snow, the weather for much of the rest of the peninsula is generally dry, with temperatures around 20°C, and winds from a north-easterly quarter. As the sun begins to return from the southern hemisphere in March and April, temperatures rise considerably, giving scorching conditions under cloudless skies.

The burst of the monsoonal air of the equatorial trough reaches the Eastern Ghats in early June and moves northwards to the Himalayas by mid-July. It arrives very spectacularly in some places: orographic ascent or great thermal heating adds to the latent instability of the airstream, triggering off torrential rain and vivid thunderstorms. But it would be wrong to imagine that the monsoon rains are equally effective in all places. Heavy precipitation is most regular in the coastal ranges of peninsular India, on the edges of the Himalayas, and in the north-east of the continent. Elsewhere, much of the rain comes from rather inconsistent migrating cyclonic disturbances, many emanating from the Bay of Bengal, and this makes the precipitation highly variable from year to year. Some areas receive very little rain, notably the Thar desert. We should also note that despite its reputation, the monsoon is not always punctual, and in some years nearly fails altogether.

The retreat of the monsoon is associated with violent storms in the Bay of Bengal and the Arabian Sea. Bangladesh has particularly suffered from these in recent years.

In summary, the climate of the Indian subcontinent can be divided into **four subseasons**: a relatively cool period from mid-December to March; a very hot dry period (April and May); the cooler wet period of the monsoon, June–September; and a warmer period during the retreat of the monsoon, accompanied by cyclonic rain.

3. Subtropical Dry Zone

Dry trade-wind air originating from the subtropical high-pressure cells controls the climate of large areas of the globe all the year round, particularly between 20 and 30° latitude. These regions of general aridity include all the popularly termed 'hot' deserts of the world: those of Sahara, Arabia, Atacama, Kalahari and central Australia are some of the major examples.

The subsiding anticyclonic circulation makes rainfall very rare in these areas, but there are perhaps only a few places where no rain falls at all: central portions of the Sahara, and also the Atacama desert in northern Chile, where no rain has been recorded for decades. Elsewhere, some rain falls in association with infrequent disturbances. In winter, lows in westerly airstreams sometimes affect the poleward margins of deserts; in summer, disturbances may occur in the easterly trades. Near high ground, local heating may set up local slope or valley winds which in unstable air cause towering clouds. Because of the intense evaporation in desert areas, raindrops are more likely to reach high ground than low ground. The Hoggar and Tibesti mountains in the middle of the Sahara desert receive appreciably more rain than the surrounding dune fields.

Because the occurrence of rain in these arid areas does depend so much on intermittent disturbances, it is inevitably extremely variable from year to year. When it does occur it is usually short-lived but torrential, causing a great deal of erosion.

With so few clouds to check incoming and outgoing radiation, temperatures reach considerable extremes. Some of the world's highest temperatures have been recorded in these arid areas, and daily maxima up to 50°C are not uncommon. The daily temperature range is also great, and can exceed 30°, especially where the moisture content of the air is particularly low. By day, the layers of the atmosphere nearest the ground can become very hazy in the extreme heat, but at night, clear skies can on occasions allow ground frost to form in the early hours of the morning.

Western Coastal Areas

On the western sea margins of many of the major deserts, a more varied weather pattern occurs, especially where a cold current occurs offshore. The cold water is brought from higher latitudes on the eastern margin of the oceanic subtropical cells (Fig. 15.3). Although the surface water becomes warmed, offshore winds from the desert set it in motion, causing it to be replaced by much cooler deeper waters. These cool the local air and frequently create fog banks which affect the adjacent coastal area.

Hence along the desert coasts of Peru, southern California, South-West Africa, Angola, and to a lesser extent Morocco, the dry warm weather of the interior degenerates into relatively cool and clammy conditions, with fog at times. Temperatures have a very low annual range. However, rainfall amounts still remain very low, because the foggy air is usually only shallow and replaced above by the more characteristic subsiding air of the subtropical belt. These narrow coastal areas sometimes have a rather specialised vegetation which has adapted itself to exist on the condensed moisture.

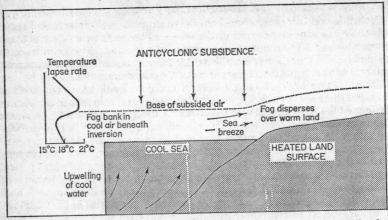

Fig. 15.3. The effect of upwelling cold water on west coast climates in the dry trade wind region (Zone 3).

4. Subtropical Winter Rain Zone

On the western side of continents, the poleward limit of the subtropical high-pressure cells lies at about 30° latitude in winter, and 40° in summer. Areas between these limits share in summer the drought conditions of the deserts at lower latitudes. However, in winter, as the high pressure retreats equatorwards, the weather of the temperate westerly belt takes over, bringing much more variable conditions, including depressions and rain. We have then, within fairly restricted areas, a well-defined type of climate with the unique quality of having its rainy season in winter and its dry season at the height of summer.

The best-known example and largest in area of this type of climate is the Mediterranean region. Here, the considerable east–west extent of sea allows the westerly influences of winter to penetrate some 2,000 miles into the Afro-Eurasian landmass. But the size of the basin and its deeply indented coastline produce many local variations on the general pattern. The eastern Mediterranean has, for example, a more extreme climate than the west. In addition, very cold winter polar air from central and northern Europe is sometimes drawn into the circulation of depressions passing through the Mediterranean; examples of such winds are the **Mistral** of the Rhône corridor, and the **Bora** at the head of the Adriatic Sea (Fig. 15.4). Southerly winds, like the **Sirocco** of North Africa, occurring ahead of depressions, are hot and dry, and can do considerable damage to crops.

Summer winds in the western Mediterranean are light and westerly, and are part of the circulation of the Azores high. In the eastern Mediterranean, they are more northerly, and are called **Etesian** winds, bringing fine refreshing weather. These winds are caused by the strong zonal pressure gradient between the Azores high and the Punjab low.

A similar, but in many ways simpler, pattern of weather recurs in the Cape region of South Africa, California and Oregon, central Chile, and in south-west Australia. If hot dry summers and mild wet winters be regarded as definitive of the 'Mediterranean' climate, then these regions are paradoxically more typical than the Mediterranean itself. Mean summer temperatures lie between 20 and 28°C, being coolest at localities close to ocean shores, where cold currents occur in some instances. In winter, mean January temperatures are around 6–10°C, above the lower limit for plant growth. Rainfall amounts in winter are largest on the poleward side of these regions, closest to the main depression tracks. Marked orographic effects cause local variations, especially in the coast ranges of California and Chile. Despite the rain, winter sunshine is abundant, and all these areas are favoured holiday regions, both in summer and winter.

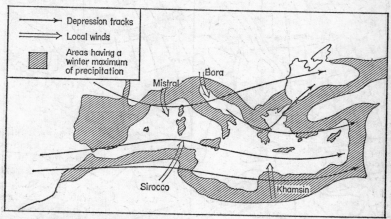

Fig. 15.4. Winds and depressions in the Mediterranean Basin in winter (Zone 4).

5. Temperate Zone

Temperate climates are found in middle latitudes and are those which come under the influence of the circulation of the westerlies belt of winds for most of the year. These climates experience both tropical and polar air masses, and as we have seen in the previous chapter, the interchange of these masses on the polar front and the vertical movement of air in the frontal systems play a major role in the weather of the whole temperate zone.

Truly temperate conditions, in the sense of being *mild* and *equable*, are found only on coastal locations on the western margins of continents. So in some senses the word 'temperate' is inappropriate, since the zone does contain some large extremes of heat and cold. One may contrast the mean annual temperature of New Orleans (20°C) with that of Winnipeg (2°C, ranging from −19° in January to +19° in July). Certainly the greatest variety of weather is found in temperate latitudes; weather assumes an importance it rarely acquires in tropical and subtropical climates. Such is its variety in some parts of the zone, the British Isles in particular, that perhaps the only reliable quality about the weather is its unreliability.

Another notable feature of the temperate zone is the strong differences in continental and marine influences from region to region, markedly so in the northern hemisphere. Ocean currents and onshore winds have an important role to play here. The prevailing south-west and west winds of the zone cause surface ocean currents to move in the same general direction. The **Gulf Stream** in the Atlantic, and the **Kuro Siwo** in the Pacific carry warm surface waters from the tropics to high latitudes. Where the prevailing winds are onshore, as in north-west Europe, there is a very marked warming effect, sustaining temperate climates to 60° N and beyond (Fig. 15.5). Some authorities regard the heating of air above warm sea surfaces in the North Atlantic and Pacific as one of the greatest known exchanges of energy on the surface of the globe.

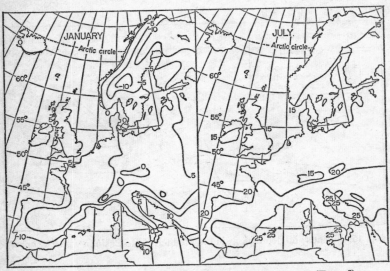

Fig. 15.5. Mean January and July isotherms over Europe (Zone 5).

In practice, two extremes of temperate climate are usually recognised: maritime, experienced on the western side of continents, and continental, characteristic of mid-latitude Eurasia and North America. The east coasts of the continents are to some extent intermediate between the two types: temperature contrasts are greater than on west coasts, but they have a more varied climate than continental interiors. This broad pattern varies in detail between Eurasia, North America and the southern hemisphere, and each of these areas will be briefly considered in turn.

Eurasia

In pressure terms, the climate of **north-west Europe** is controlled by the gradient between the Azores high and the Icelandic low, with the addition in winter of the Siberian high. In winter, when the direct influence of the Azores high is at its weakest, the average strength of the westerlies is about twice that in summer. So although maritime Europe is affected all year by a succession of fronts, depressions and anticyclones, weather changes in summer are less marked. The longest spells of settled weather come from blocking anticyclones. Despite the seemingly random pattern of weather changes in Europe, there is some statistical support for certain beliefs, long cherished in folklore, that at certain times of the year similar weather sometimes recurs. In Britain, the second week in September has a reputation for anticyclonic weather. These recurrences are referred to as **singularities.**

The warmth of the Gulf Stream and its continuation, the North Atlantic drift, ensures that mean winter temperatures in north-western Europe are generally above 0°C (see Fig. 15.5). In coastal Norway, this situation persists as far north as 70° latitude, where the mean temperature for January is 28°C above the average for the latitude. In summer, mean summer temperatures in maritime Europe lie between 15 and 16°C.

There is no dry season in north-western Europe, although the month for maximum precipitation varies: in western Britain and other Atlantic situations, it is an autumn or winter month, when depressions are at their most vigorous. In south-east England, and throughout central and eastern Europe, summer is the time of maximum rainfall, although the frequency of wet days is actually less in summer than in winter. However, the most marked precipitation differences in Europe are largely created because of topography. The frequency of precipitation in north-west Spain, Wales, the Lake District of England, Scotland and western Norway rises to over 200 days a year, with an annual precipitation of up to 5 m (*c.* 200 in.) in the most exposed situations. This compares with 167 rain days in London (Kew), producing an annual precipitation of only 0·6 m (24 in.).

Further eastwards through central and eastern Europe into **Russia**, continental air masses from the east play an increasing part in the climate. The absence of marked topographic barriers makes the transition a very gradual one, and westerly depressions regularly penetrate well into the heart of the continent. In the region of Moscow (56°N), where the mean January temperature is −10°C, periods of rain with melting snow from westerly depressions alternate with icy snowstorms and temperatures of −35°C. The amount and frequency of precipitation decreases gradually towards the interior of Asia. Where mountains interfere, westerly influences can be excluded almost altogether, and precipitation amounts made very low. The Tarim basin in east Turkestan (40°N) has been described as the most continental region on earth.

In the temperate zone of southern Russia, July temperatures are above 15°C; this is also the period of maximum rainfall from depressions and convectional activity. On the other hand, in winter, temperatures are well below freezing. Although westerly airflow continues in the upper air, the dominance of the surface Siberian anticyclone centred further north is strong.

Moving further east still, westerly winds again play a prominent role in the climates of **China** and **Japan**. Westerly depressions regenerate in the lee of the Tibetan plateau, bringing substantial precipitation along the main frontal zone. In winter, this lies over southern China and to the south of the main Japanese islands. North of this line, cold air from Siberia occupies much of China, and can bring freezing conditions a long way south. This air flow rapidly transforms over sea areas to create heavy snowfalls in the Japanese mountains. The frontal zone moves northwards into northern China in June and July, the time of the so-called south-east monsoon. In reality, this is a very shallow wind system at these latitudes and has little effect on the weather. Summer rain in China comes mainly from the south-west or south. The retracing of the polar front southwards brings intense summer rains, and at this time of the year typhoons are frequent in the China Sea.

North America

A general resemblance to the climate of north-western Europe can be found in **British Columbia** and the northern **Pacific States**. However, one major factor intervenes: the extensive western Cordillera strongly discourages the eastward penetration of maritime air masses into the heart of the continent. In addition, the shape of the coast prevents the drift of warm waters into latitudes as high as in the Atlantic. The net result is that a climate of the temperate maritime type is more restricted in extent than in Europe. But as in Europe, topographic factors make the climate variable over short distances: a regular pattern of rainy windward slopes alternating with drier lee slopes obtains in the north–south ranges. Some interior valleys require irrigation for crops.

In the states and provinces of the **interior** and the **east coast**, unlike Eurasia, there are no high east–west mountain barriers to prevent the free interchange of cold and warm air masses over the whole temperate zone from the Gulf of Mexico to the tundra margins (Fig. 15.6). Very cold air from the Mackenzie anticyclone may penetrate to the Gulf Coast on several occasions each winter, producing temperatures as low as −10°C and causing considerable damage to crops. At other times, temperature changes further north from −15° to +15°C

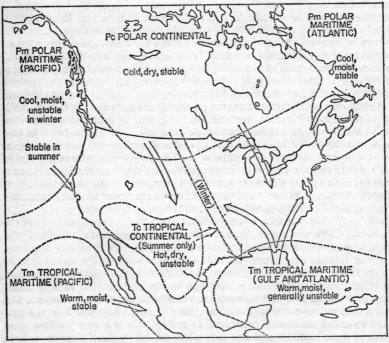

Fig. 15.6. Air mass centres and air flows over North America.

are not unusual in the space of 24 hours. However, generally speaking, the winter dominance of cold polar air gives remarkably low average temperatures in the interior and on the east coast. In Omaha (41°N) the January mean is −6°C; in New York it is −1°C. This is rather different from a typical figure of 8–9°C for a Mediterranean station at the same latitude.

Conversely, in summer, tropical air masses from the Gulf of Mexico bring a moist sweltering heat to the southern States, the east coast and the Great Lakes region. Mean summer temperatures at Washington D.C. are +26°C, which is close to the mean annual temperature of the equatorial zone.

For all this, we must not lose sight of the fact that the interior and east coast areas are under the influence of the zonal westerlies throughout the year. Major depression tracks trend west–east across the continent, many originating in the lee of the western mountain ranges in Texas and Alberta. These depressions frequently interject periods of disturbed weather into the seasonal patterns, and prevent the development of a climatic type of the Central Asian type, which has little winter precipitation.

A major region of cyclonic activity lies off the coast of New England in winter. Here, polar air masses passing over the cold Labrador current meet warm air from the south, along the North Atlantic Polar Front. However, the prevailing westerly airflow prevents the oceanic influence from being very strong on the Atlantic coasts of America, unlike Europe. The Great Lakes sometimes have more influence: they can cause westerly airstreams to pick up moisture and give heavy snowfalls on south and eastern shores. In summer, the Lakes have a slight moderating influence on temperatures.

Southern Hemisphere

The great westerly winds of the southern hemisphere travel almost entirely over open ocean, and not surprisingly air masses are strongly maritime in

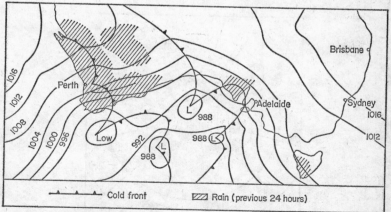

Fig. 15.7. Complex winter depression over southern Australia, bringing rain to Perth and Adelaide regions (Zone 4). In summer, the main depression tracks move southwards, and only Tasmania receives cyclonic rain all year (Zone 5).

character. Particularly noticeable is the evenness of seasonal temperature and the smallness of variations in the weather from season to season.

For equivalent latitude, temperatures are much lower than in the northern hemisphere, particularly in the south Atlantic and Indian ocean regions, where Antarctic icebergs reach into middle latitudes. Bouvet Island (55°S, 3°E) is covered with glaciers almost to the sea. At these latitudes, 300 days of rain a year and a cloud cover of 80–85 per cent are quite common.

To seek a climate equivalent to temperate Europe or America, we find it between 40° and 45°S in Chile, Tasmania and New Zealand. The South Island of New Zealand lies just to the north of the main active convergence zone in the South Pacific and rainfall occurs at all seasons. A marked rain-shadow effect exists in the lee of the South Island Alps in the Canterbury area. Similar contrasts exist in Tasmania and especially in Chile, where the Andes help to sustain a virtual desert on the Patagonian side of the range.

6. Subpolar Zone

Beyond about latitude 60° in the northern hemisphere and 55° in the southern, we move into a zone of rather inhospitable climate where incursions of warm air masses become very much rarer than in the temperate zone. Although the influence of the westerlies is still felt, the effects of the polar and northern continental anticyclones increasingly dominate the weather with higher latitude. The seasons become emphasised by the great difference between summer and winter daylight hours. In the northern hemisphere this climatic regime includes the northern part of North America from Alaska to Labrador; southern Greenland and Iceland; northern Scandinavia and Siberia (Fig. 15.8).

In Siberia and Arctic Canada, we find a climate whose main characteristics are a great seasonal temperature range, extremely severe winters, and a small annual precipitation concentrated in the summer months. Winter is the dominant season, lasting 7–9 months. Typical mean January temperatures

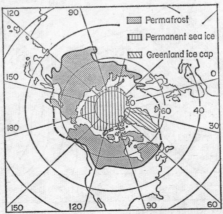

Fig. 15.8. Permafrost, sea-ice and ice-cap distribution around the North Pole (Zones 6 and 7).

over much of these two large areas lie in the range −20 to −40°C. Some exceptional figures have been recorded in low-lying situations: Oimekon (60°N, 140°E), in the Lena valley, has experienced a temperature of −78°C, the coldest in the northern hemisphere. Under the influence of the intense high pressure, winters are generally calm, with only light precipitation.

Summers are short, with intense insolation, but the weather changes frequently, bringing heavy rain. The length of summer is not usually sufficient to thaw out the frozen ground of winter to more than a metre or so, and permafrost persists beneath. In eastern Siberia permafrost reaches as far south as 50°N. However, high diurnal temperatures (20°C) do allow coniferous forests to flourish in some sheltered valleys.

The marked seasonal contrasts of these continental regions produce enormous annual temperature ranges, reaching over 60°C in Siberia, the largest recorded anywhere on the surface of the globe.

In Iceland and southern Greenland, and other oceanic situations in the subpolar zone, the climatic extremes of the continents are not experienced. Instead, a rather damp and cold regime persists all year round. Storms from varying directions occur frequently; clouds are thick, and rain, snow and sleet fall in quick succession on 250–300 days a year. Low summer temperatures of the order of 10°C and strong winds create an almost completely treeless landscape, although the ground flora can be quite rich. Winter temperatures are usually just below freezing point. The high precipitation amounts nourish ice-sheets and glaciers in the higher mountains surrounding the North Atlantic and Pacific Oceans.

Where the northern land-masses reach the Arctic Ocean, the temperature contrast between sea and land is sometimes sufficient to initiate unstable frontal activity. This is the normal position of the Arctic Front, which is at its most pronounced in the north-slope region of Alaska, generating intense east-moving cyclones and much bad weather.

In the southern hemisphere, the subantarctic zone is stormier and very much colder than the oceanic regions of the northern hemisphere. Unlike the northern hemisphere, the circulation is predominantly zonal—that is, winds and cyclones come from the east at these latitudes. Temperature differences between places on the same parallel of latitude are minimal, again in marked contrast to the other hemisphere. There are only a few land stations in this zone: the Falkland Islands, South Georgia, the southernmost tip of South America, and some other small islands.

7. High Polar Zone

Antarctica, Greenland and the Arctic Ocean, by far the largest three regions of ice that exist on the globe, have their own characteristic climatic type. In common, they differ from the subpolar zone by possessing lower average temperatures, fewer depressions, and by lying entirely within easterly zonal circulation. A considerable amount of new climatic data has been collected in these regions, especially in Antarctica, since the 1957 International Geophysical Year.

The **Arctic Ocean** has a variable climate, but with less frequent strong winds than over the open ice-free ocean further south. The peak of summer, May to July, brings much quiet but overcast weather, with temperatures just above freezing point. The process of melting in summer takes up so much latent heat that temperatures are prevented from rising any higher. Autumn is the most disturbed season with snow and gales. Quiet weather prevails in winter with clear skies; temperatures lie a little below −30°C, which is somewhat less severe than on adjacent continental land masses.

In **Greenland** and **Antarctica**, the ice-caps are nourished by cyclonic precipitation. However, snowfall is not evenly distributed. In Greenland the heaviest precipitation is received in the southern part of the cap, which rises to 2700 m. North Greenland is a cold desert and receives very little snow. In Antarctica, the average annual increase of snow cover for the whole continent is about 15 cm, but snowfall is only light near the pole.

Temperatures near the centre of Greenland appear to have a maximum in summer of about −10°C; February is the coldest month (−47°C). However, Antarctica is the most intense of all these cold centres. Summer temperatures in the interior remain below −20°C, although they may rise close to freezing point around the fringes of the continent. In winter, temperatures near the pole regularly fall below −70°C. The Russian meteorological station Vostok has recorded the world's lowest surface air temperature of −88°C. This occurred towards the end of the long winter night.

In both Greenland and Antarctica, the intensely cold air over the ice at times flows downslope under the influence of gravity towards the margins. This flow can gather momentum to create severe blizzard winds which funnel down valley glaciers towards the coastal margins.

Suggested Further Reading

Flohn, H., *Climate and Weather* (Chapter 5), Weidenfield & Nicolson, London, 1969.
Hare, F. K., 'The concept of climate', *Geography*, 1966, **51**, 99–110.
Miller, A. A., *Climatology* (6th edn.), Methuen, London, 1953.
Trewartha, G. T., *The Earth's Problem Climates*, McGraw-Hill, New York, 1961.
Trewartha, G. T., *An Introduction to Climate* (4th edn.), McGraw-Hill, New York, 1968.

CHAPTER SIXTEEN

LOCAL CLIMATE

There are many significant controls of climate which operate on far less grand a scale than those determining the major climatic zones of the world. Important local climatic variations can be set up by features ranging from mountain barriers and large lakes to small differences in the nature of the ground surface. Local factors like these often go a long way to explain the apparent differences between the weather we hear forecast on the radio and television, and that which we actually experience. Outside mountain areas, local contrasts are often seen best developed during calm episodes in the general circulation. In mountain areas, important variations exist at all times.

The Climatic Effects of Relief

As one climbs up a mountain, it is a well-known experience that changes take place in several of the major climatic elements, particularly in temperature and pressure. The causes of these changes have been discussed in earlier chapters. Here we can concentrate on the effect of the shape of relief on local climate. This effect operates through its influence on the flow of air and in creating temperature contrasts.

Aspect

Aspect can have a bearing at all scales on temperature in local climates. A south-facing garden receives higher insolation amounts and is likely to be much warmer than one facing north on the shady side of a house. These insolation contrasts become much more acute in mountain regions because of the general higher intensity of radiation in the thinner atmosphere.

In the Alps of Europe, large mountain valleys running east–west are noted for having a south-facing side which is well endowed with sunshine, and a north-facing side which may be in constant shadow for much of the winter. Not only will this result in strong day-time temperature contrasts, but the entire economy of the valley may well be adjusted to it: farms and villages occupy the sunny side, while the shady side is left under forest (Fig. 16.1).

At similar latitudes in the western United States, a different pattern results. Thick forests grow on north-facing slopes, but the opposite side is often bare, or supports only cactus and a few weeds. In this case, we are dealing with areas of low rainfall: moisture is conserved sufficiently on shaded slopes for forest growth, but there is a considerable water deficiency on the heated south-facing slopes. This example serves as a warning that the well-known Alpine case should not be taken as a universal rule.

Precipitation

The general rule is that on the windward slopes of high ground, the orographic effect tends to add to whatever other rain-giving mechanism may be in operation. Even stable air may be induced to part with some moisture. With

189

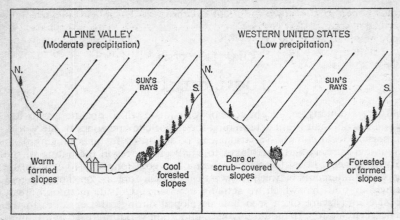

Fig. 16.1. The effect of aspect on an east–west valley.

conditionally or potentially unstable air, mountains provide the stimulus for the release of often heavy precipitation. The heaviest rainfall occurs when warm moisture-laden air is forced over mountains, as in the summer monsoon of India. The world's highest precipitation totals occur on the windward side of tropical mountains: Cherrapunji (north-east India) receives more than 10 metres (400 in.) of rain per year, and areas in Hawaii almost 12·5 metres (500 in.).

On the lee side of mountains, the relative deficit of precipitation creates a rain-shadow. The air, having lost much of its moisture, is now descending and becoming effectively drier. Even frontal rain may be interrupted, giving a temporary clearance of cloud. Thus by and large, as a rainfall map will confirm, the local distribution of rainfall in any one area often reflects the relief of the ground. We have come across some examples of this in the last chapter: sharp climatic divides can occur within very short distances, as among the ranges of the Pacific north-west of the United States.

These general points do not apply quite so readily where the precipitation is largely snow. Although much of the snow may initially fall from clouds on the windward sides of hills or mountains, a considerable amount may be blown over the hill crests to actually accumulate on the lee side.

Air Flow

Any obstacle can markedly affect the smooth flow of wind, be it a house, or a belt of trees, or a large mountain. The degree of effect varies as much with the shape of the obstacle as with its size, as well as with the wind speed. When wind approaches a relatively large object like a hill, there is a forced inducement to uplift (orographic uplift). The uplift may be reinforced by thermal influences, especially near south-facing slopes, and clouds will form, depending on the stability and moisture content of the air.

Many of the more noticeable effects are likely to occur downwind of the high ground. The windy conditions of the summit will give way to relatively calm conditions on the lee side, or in the case of a large mountain, the flanks may be subject to a downwind. This may produce the type of Føhn effect described in Chapter Eleven. Further away from the hill, marked turbulence and eddying is likely to occur, and strong wind conditions may produce a rotor circulation, sometimes revealed by a roll cloud which remains stationary in the lee of the hill whilst the air flow spirals through it. In conditions of less disturbed flow, lee waves may develop (see Fig. 11.7).

In addition to the direct mechanical effects of relief on air flow, elevated areas can create special local winds because of the temperature contrasts between the valley bottoms and the higher slopes. Winds created in this fashion are given the general term **mountain and valley winds**.

During the day, providing conditions are fairly calm and skies relatively cloud-free, warm air tends to blow up the valley in response to the heating of air in contact with the upper slopes of the valley and the surrounding upland. A return wind aloft completes the circulation (see Fig. 16.2). Similar circulations on a smaller scale are likely to be in force on the actual slopes of the valley. Such upslope and upvalley winds are described as **anabatic**. At night, the situation reverses: the upper slopes cool more quickly and dense cold air drains downslope towards the valley bottoms. This night-time or **katabatic** wind is generally stronger than its day-time counterpart and has more noticeable effects.

The accumulation of chilled air in valley bottoms and sheltered hollows can lead to the development of severe frosts, with obvious hazards to local agri-

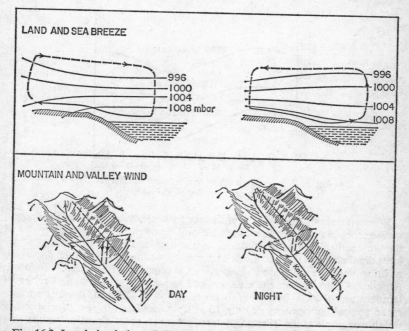

Fig. 16.2. Local circulations (*left*, day; *right*, night): (*top*) land and sea breezes and (*bottom*) mountain and valley winds.

culture. Where this happens frequently, the low-lying ground may have the reputation of being a **frost hollow**. Under certain circumstances, very low temperatures can occur in these: the coldest known spot in Europe is a frost hollow known as the *Gsteltneralm*, a deep limestone sink-hole in the Austrian Alps. Temperature differences of the order of 30°C have been recorded between rim and bottom.

Frost-hollow effects can also be felt on a very modest scale, as in sheltered valleys in south-east England, or even in small hollows in fields and gardens. Hence minor differences of topography can have a considerable bearing on frost-sensitive activities such as market gardening and vine cultivation.

Mountain Climates

With all these local effects, it is perhaps not surprising that the mountain areas of the world do not fit neatly into any scheme of major climatic zones. They are really an amalgam of many climatic types which change rapidly over short distances, both horizontally and vertically. In a general way, the rise in altitude in mountain zones is equivalent to an increase in latitude; for instance, the equivalents of subpolar and polar climatic types are found among the snowfields above the treeline. One also finds a related succession of vegetation life zones as one ascends a mountain. The general analogy between elevation and latitude is not perfect, however, because of the divergence in insolation patterns.

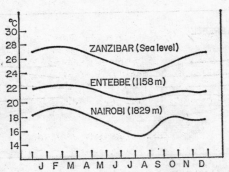

Fig. 16.3. Annual march of temperature in East Africa,
showing the effect of altitude.

Where mountain areas are extensive, or form elevated plateaus, the drop in temperature is perhaps the most significant contrast with lowland regions at the same latitude. In equatorial regions, most upland areas still maintain their latitudinal characteristic of a small temperature range, but at much depressed mean values. The situation in East Africa is illustrated in Fig. 16.3. In South America, Quito lies on the equator and has a very small annual range of temperature (0·4°C), but being at an elevation of 2850 m, it experiences a mean annual temperature of only 12·5° C. Away from the equator, another notable example is the Tibetan plateau, where its great elevation results in a climate more akin to Siberia than to an adjacent region such as southern China.

Individual mountain peaks or isolated ranges often form relatively humid zones in much drier surrounding lowlands. Even in desert regions, mountains normally receive some precipitation. This renders them of great economic importance in many countries as water storage regions.

The Climatic Effects of Water Bodies

On a broad scale, we have seen that oceanic climates are milder than continental climates at the same latitude. Under the influence of winds from the sea, winter temperatures are higher and summer temperatures cooler. These general effects are reinforced in a local context round the immediate margins of sea or lakes, creating in effect a recognisable type of coastal climate. Coastal areas have the smallest range of temperatures. This is reflected in many situations by the low likelihood of frost or snow. On the other hand, coastlines frequently suffer from fog: in summer, warm air moving over cold water is the predominant cause (advection fog); in autumn and winter, fog often occurs because of the cooling of moist sea air as it moves inland.

Another typical local effect in coastal areas occurs in the form of the **land and sea breeze** mechanism (see Fig. 16.2). On a warm day, air over the land becomes heated and expands more than the air over the relatively cool sea. The expansion of the air column over the land tilts the local pressure gradient landwards near the ground surface, and seawards higher up. The net result is that a small circulatory system is set up in which a sea breeze blows landwards during the day, and is compensated by outblowing winds aloft. At night, the same principles apply in the reverse situation: the air over the sea is warmer, and an offshore land-breeze results.

These breezes can have a marked effect on coastal climate. The day-time sea breeze usually causes both a drop in temperature and increased humidity relative to areas further inland. In middle latitudes, the Coriolis effect gradually causes the sea breeze to be deflected, so that by late afternoon it eventually blows parallel to the shore. Similar features to land and sea breezes can be observed around the shores of large lakes, notably the Great Lakes of North America.

The Climatic Effects of Vegetation

Most plant life is found within the first 2 m (6 ft) of the ground, and nearly all of it within the first 100 m (300 ft); within this zone, some distinct local effects are noticeable. All vegetation, to a greater or lesser extent depending on its structure and density, protects the underlying surface from the temperature extremes that can be observed over bare ground. The degree of modification becomes especially marked where the vegetation is in *forest* form. Here, the local energy budget, winds, air flow and humidity are all changed.

Forest Climates

A well-developed forest canopy blankets both incoming and outgoing radiation. The radiating surface is transferred from the ground to the treetops, which therefore manages to intercept nearly all the sunlight. The highest temperatures consequently occur at the top of the canopy. The proportion of radiation reflected or trapped depends to some extent on the type of tree: dark-leaved trees absorb much more energy than young, light-coloured specimens, which have a higher albedo effect.

The amount of light getting through the canopy varies enormously from forest to forest: about 50–75 per cent may reach the floor of a birch/beech forest, 10–25 per cent with spruce and fir, and as low as 0·1 per cent in a tropical forest. More light obviously penetrates deciduous forests when they are leafless. Temperatures inside the forest are generally lower by day but higher by night compared with areas outside. The day-time difference is of the order of 3°C in temperate latitudes. An additional factor is that temperatures inside forests are also depressed by evapotranspiration from the plants. At night, where the forest canopy is well developed, the tree crowns are usually the coldest part of the forest, but with less dense woods the cold air can sink to give a temperature minimum near the ground.

Another climatic characteristic of forests is that they show humidity increases of up to 10 per cent compared with open ground; this is especially so in summer. Measurements taken in central Europe show that the increase varies with the type of forest: for beech it is 9·4 per cent; for larch, 7·9 per cent; and for Scots pine, 3·9 per cent. Two factors contribute to making the forest environment more humid: the moisture from the trees, and the lack of evaporation loss within the forest because of lower temperatures and wind velocities.

Finally, forests give a certain amount of shelter from wind and rain. The interception of precipitation depends on the density of the crown and the type of rain. Drizzle is much more totally intercepted than heavy downpours. The intercepted precipitation either evaporates on the canopy, drips to the ground or runs down the trunk.

Air movement in forests is slight compared with the open, and quite large variations in outside wind velocity have little effect inside woods. Fig. 16.4 represents some of the measurements taken of the progressive reduction of wind speed in a deciduous forest. In a tropical forest the reduction would be even more marked. One of the side-effects of wind-speed reduction is the increased possibility of frosts in clearings. Knowledge of the influence of forest trees on wind has been useful in the construction of artificial windbreaks to protect crops and homes. Too effective a windbreak may lead to acute turbu-

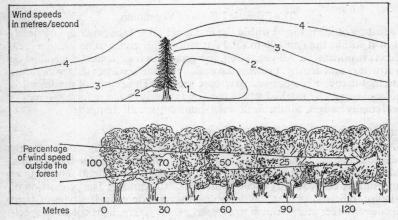

Fig. 16.4. Wind penetration of a forest and shelter belt.

lence downwind of the sheltered zone. The best general protection is given by a barrier allowing about 40 per cent penetrability.

Other Vegetation

All vegetation with horizontally-spreading leaf canopies, from garden flowers to small clumps of trees, exhibit some of the characteristics of forest climates, albeit on a microscale. On the other hand, vegetation with a more vertical structure, such as standing wheat, barley or maize, allows the warmth of the sun to penetrate more effectively down towards the ground. In addition, air is trapped between the many stalks and is not changed by turbulence. This means that high temperatures can develop within the crop, and at night the cooling effect is also distributed downwards within the plants. Hence vertically structured vegetation is more prone to extreme temperatures.

In the case of a **grass sward**, day temperatures are kept comparatively low by the cooling due to evapotranspiration, and at night the moist stagnant air between the grass stalks is additionally cooled by radiation. We find that low temperatures and frost are especially likely over grassy surfaces and more so over long grass than short. Heavy dew is most common on long grass. Sometimes ground fog may develop, usually less than knee-high deep, and often seen over river meadows in the early evening, perhaps as a precursor to a more general radiation fog later in the night.

Urban Climates

The city-dweller, no less than those who live in mountain areas or within forests, experiences a climate distinctive from the regional pattern. The agglomeration of buildings interferes with the wind and atmospheric characteristics to a degree at least equal to that of a large forest.

As within forests, urban structures tend to reduce wind speeds to lower values than those recorded in open country. But the effect varies with the street pattern, the time of day and season. The wind tends to channel down streets parallel to the general direction of flow, especially in a city with canyon-like streets, as in New York. Conversely, in streets running at right angles to the wind, strong lee effects may be experienced. During the day, city wind speeds are considerably less than surrounding areas, but at night, turbulence over the city makes contrasts less apparent. Rural–urban contrasts are most marked with strong winds, and the effects are therefore more evident in winter than in summer.

Unlike forests, cities tend to have lower humidities than their surroundings: the general absence of vegetation and large bodies of water, and the rapid removal of surface run-off, all contribute to decrease local evaporation. On the other hand, it seems likely that under certain conditions, thermal and turbulent influences over cities may trigger off precipitation or thunderstorms. In Europe and North America, cities generally record 11–18 per cent more days with light rain and thunder than surrounding country, resulting in a slight increase in total precipitation.

The Heat Island Effect

Urban areas are generally warmer than the surrounding countryside. There are three main factors responsible for this: the direct production of heat in the city from fires, industry and central-heating systems; the heat-conserving properties of the brickwork and fabric of the city; and the blanketing effect by atmospheric pollution on outgoing radiation. The centre of London has a mean annual temperature of 11·0°C, compared with 10·3°C for the suburbs (Kew) and 9·6°C for the surrounding countryside. The contrasts between maximum temperatures are small; the biggest difference is in minimum temperatures: there are very few frosts in Central London compared with outer areas.

Measurements taken throughout the year in London and other cities have shown that the greatest heat contrasts between town and country occur in the summer, especially at night after a fine sunny day. Since at this time of the year direct heat from combustion and atmospheric pollution are at a minimum, we must conclude that the strongest factor at work creating a heat island over the city is the storage of heat from insolation in the brickwork. This heat island operates most effectively when wind speeds are low. In the high insolation conditions of autumn and summer, on a calm night the heat island has steep thermal margins, and the highest temperatures are clearly seen to be associated with the highest density of buildings and road surfaces (Fig. 16.5).

Atmospheric Pollution over Cities

City atmospheres are notoriously liable to pollution, affected by soot, ash, sulphur dioxide, gases, fumes and smoke. These have the effect of blanketing

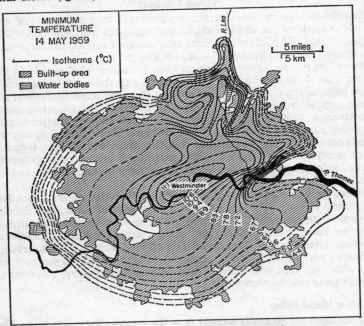

Fig. 16.5. The 'heat island' effect over London.
(From T. J. Chandler, *The Climate of London*, Hutchinson)

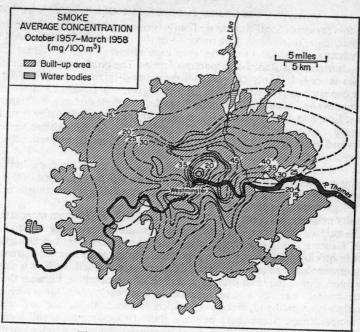

Fig. 16.6. Smoke pollution over London.
(From T. J. Chandler, *The Climate of London*, Hutchinson)

the radiation over a city, cutting down the sunlight, and providing abundant condensation nuclei. Under normally windy conditions, much of this atmospheric waste is diffused upwards by turbulence and removed by stronger winds at height. The greatest concentrations of smoke occur with low wind speeds, temperature inversions and high relative humidities.

Some British cities lose 25–55 per cent of incoming solar radiation in the winter months; December is particularly bad, when the insolation strikes the pollution layer at a low angle. Poor sunshine records may in some cases also be a consequence of the high frequency of fogs, which, given appropriate temperatures, readily form because of the abundance of condensation nuclei. Occasionally, under stable anticyclonic conditions, radiation fog combines with excessive pollution and becomes trapped under a temperature inversion to form **smog**. One of the most unhealthy of these occurred in London in December 1952, resulting in many more deaths than usual for the time of year. In Los Angeles, smogs form in summer and autumn, and are largely caused by the concentration of pollution from car exhausts under the daytime subsidence of air which creates the necessary temperature inversion.

The fact that London now rarely suffers from a smog says much for the effectiveness of measures taken since 1956, the date of the Clean Air Act. More will be said of the control and conservation of the environment in Part Four of this book.

The Climate Nearest the Ground

In this review of local climate we finally turn our attention to climate on the smallest scale, that within a metre or so of the ground. It has been well said that climate often varies more between man's head and his toes than it does over a horizontal distance of hundreds of miles. The outstanding feature of the surface layer is its daily temperature range, and this can vary considerably from one surface type to another. In a town, one can readily recall instances where snow or frost vanishes rapidly from dark asphalt surfaces, but lingers on lawns. Likewise, in the country, some types of field appear to suffer from frost far more than others. The colour of the surface, and the relative proportion of solids and water in it, are probably the two most important properties to cause the variations.

At one extreme, **bare rock** or built surfaces may warm up considerably in sunshine, more so if they are dark rather than light in colour. Some of this heat is stored and re-radiated at night, maintaining a relatively warm surface, as in the brickwork and asphalt in towns.

A **dry or sandy soil** containing a good deal of air heats up even more strongly at the surface than rock during the day, because air is a poor conductor of heat. Hence a sandy beach can become very hot to the feet; but at night this heat is quickly lost. Sandy soils therefore display great diurnal extremes of temperature and are prone to frost. Conversely, a **wet soil**, such as a clay soil in winter, shows the least diurnal variation of temperature, because the high water content ensures a slow reaction to heat exchange. But a wet soil is usually also a cold soil, since evaporation lowers both surface and adjacent air temperatures. Somewhere between these two extremes, a soil such as a medium loam has moderate thermal properties, being generally warm and free from sharp temperature fluctuations.

Snow presents the coldest surface of all. During the day a large proportion of the incoming solar radiation is reflected away and there is little gain of heat. At night, heat loss by long-wave radiation proceeds very rapidly and very low minimum temperatures may be recorded just above the snow surface. However, snow is about 90 per cent air and forms a very effective insulating blanket, preventing temperatures underneath at the ground surface from falling too far below freezing. Sheep, which would otherwise have perished in the open, have been known to survive the severest blizzards by being buried under snow.

The variations in climate near the ground become less apparent under windy conditions, and where the ground is covered by trees or buildings the local climate created by them tends to blur some of the effects of the different ground surfaces.

Suggested Further Reading

Chandler, T. J., *The Climate of London*, Hutchinson, London, 1965.

Flohn, H., 'Local wind systems', in *World Survey of Climatology* (Vol. 2), Elsevier, Amsterdam, 1969.

Garnett, A., 'Insolation, topography, and settlement in the Alps', *Geographical Review*, 1935, **25**, 601, 617.

Geiger, R., *The Climate near the Ground* (4th edn.), Harvard University Press, Harvard, 1966.

Manley, G., *Climate and the British Scene*, Collins, London, 1952.

CHAPTER SEVENTEEN

CLIMATIC CHANGE

One often hears that 'the climate isn't what it used to be'. Nostalgic recall of long Edwardian summers or the perpetual clear skies of the Battle of Britain during the Second World War inevitably raises the question: is our climate constant or does it change with time? This topic is not only a subject for considerable speculation, but has also given rise to a large body of scientific research, especially in recent years as our better understanding of the workings of the general circulation has made it possible to identify some of the more immediate causes of climatic change.

Two main aspects of climatic change will concern us in this chapter, that of climatic record and the possible causes of change. The evidence for reconstructing the record is very diverse, covering archaeology, history, botany, zoology, geology and glaciology, as well as meteorology and oceanography.

The Climatic Record

The primary evidence for climatic change many millions of years ago is the rocks themselves. Their sediments and fossils tell us a great deal about the environment in which they were laid down. We may find in close proximity to one another, coal deposits indicating the humid conditions of the tropics, red sandstones laid down in deserts and morainic material reflecting polar conditions. We must, of course, bear in mind when interpreting this evidence that, because of the movement of the continents, no one latitude of the Earth has necessarily gone through such marked climatic vicissitudes, because any one part of the crust may have wandered through several climatic zones during its geological history.

Nevertheless, even allowing for this factor, one of the most remarkable discoveries about the climate of the past is that the two poles of the Earth, whatever the distribution of continents in relation to them, have been free of ice for at least 90 per cent of the known history of the planet. In other words, despite the clear legacy in many parts of the world of recent glaciation, the vast majority of sedimentary rocks were laid down in warm climatic conditions which appear to have been relatively uniform over large stretches of the Earth's surface.

Ice Ages

At intervals of many tens of millions of years, geological history has been punctuated by five or six glacial episodes, the oldest known being about 2500 million years ago. The last three occurred at the beginning of the Cambrian (550 million years ago); during the late Carboniferous and Permian periods (250 m.y.); and in the Pleistocene (the last 2 m.y.). Some of the spatial effects of Pleistocene glaciation are described in Chapter Six. Any theory which tries to explain the cause of ice ages would need to take into account not only the tremendous gaps of time between each ice age (macro-scale variations of climate), but also the glacial/interglacial oscillations within them (meso-scale variations). We have seen that during the Pleistocene, at least seven or eight warm/cold cycles have been recorded (see Table 19.1).

At the time of each ice advance, the climate of the whole world appears to have been affected, although there is considerable debate about whether the changes were entirely synchronous throughout the globe. The major climatic belts changed their extent or position, resulting in temperatures about 5°C lower than now in the tropics, and up to 15°C lower in the glaciated middle latitudes. The spectacular nature of these changes inevitably raises the questions, what causes an ice age, and are we to expect another one? We can try to answer this question in a general discussion of causes later in the chapter.

Flandrian Postglacial Changes

With the final retreat of the glaciers around 10,000 years ago (8000 B.C.) the climate rapidly ameliorated in middle and higher latitudes. A thermal maximum some 2–3°C warmer than now was reached between 5000 and 3000 B.C. This period is contemporaneous with a wet sub-pluvial episode in Australia and North Africa; settlements flourished in the Sahara. At about 500 B.C. the climate of Europe deteriorated, and a cool wet period set in.

Much of the evidence for this broad pattern of change comes from the study of the remains of plants and their pollen. The Scandinavian botanists A. Blytt and R. Sernander have put forward a more detailed scheme based on their work on ancient tree layers and peat deposits (Table 17.1). The names suggested by them, notably **Boreal** and **Atlantic**, are in common use as the main divisions of the post-glacial period, although their precise reality in climatic terms is still a matter for discussion.

Table 17.1. Postglacial Climatic Stages

Years B.C.	Stages	Climate	
500–present	Sub-Atlantic	Cool and wet	
3000–500	Sub-Boreal	Warm and dry	Climatic optimum
5000–3000	Atlantic	Warm and wet	
7500–5000	Boreal	Cool and dry	
8300–7500	Pre-Boreal	Cold	

Source: After A. Blytt and R. Sernander.

Climatic Change in Historical Times

Although it needs care in its interpretation, a great deal of information, mainly literary, is available for the climate of the last 2000 years or so. But this written material is not in the form of accurate meteorological observations. Instead, we have to rely on such works as contemporary descriptions of the weather, which inevitably tend to emphasise storms and floods; agricultural records, particularly the dates of harvesting and crop yields; the records of port closures and openings each winter in northern Europe; and sagas of voyages and colonisations. Even the paintings of Constable have been looked at for clues to contemporary weather. From all this, we glean a fascinating tale of climatic variability which in a time of largely rural societies had an important bearing on everyday livelihood.

After the severe deterioration of climate at about 500 B.C., several centuries of poor conditions in northern Europe were followed by drier conditions in the Roman era. This change also seems to have affected much of southern Europe, Asia and North Africa. On the southern shores of the Mediterranean, Roman settlements such as at Carthage increasingly suffered from problems of water-supply, as witnessed by the building of aqueducts. The increasing drought in Asia has been suggested as being an underlying cause of Barbarian invasions of Europe.

The period A.D. 400–1200 was on the whole dry and warm and apparently remarkably storm-free in the Atlantic and North Sea. This was the time of the great Viking voyages to Iceland and Greenland, and possibly America. Grapes were widely grown in England, implying summer temperatures perhaps 1 to 2°C higher than now. But after 1200, another period of weather variation and general decline set in. A number of devastating floods and storms are recorded in northern Europe around 1300. In Greenland and Iceland, the deterioration was particularly marked, and the Viking colonies virtually froze to death. Despite a partial recovery 1400–1500, when southern fruits were introduced into English gardens, conditions remained relatively cool in the sixteenth and seventeenth centuries. From 1540, and still more frequently after 1600, ice blocked the coasts of Iceland for an average of 5–6 months each year.

The period 1550 to 1800 has been called 'the Little Ice Age', because in it the glaciers of the mountains of Europe reached their most advanced positions since the beginning of the postglacial epoch. This advance has left well-marked terminal moraines, known collectively as the **neoglacial maximum limit**. H. H. Lamb has suggested that polar sea ice at this time stretched south almost to the Faroes and the Shetland Islands (Fig. 17.1). In Britain, interesting evidence is provided by the record of the freezing over of the Thames for each century (Table 17.2). Records cease to be comparable in the nineteenth century because the new London Bridge built in 1831 allowed free tidal movement of water in and out of the river.

The Little Ice Age was a period of agrarian distress in northern countries, farmland having to be abandoned to the ice in Norway, Iceland and the Alps. For the 1780s, we have sufficient evidence to conclude that average temperatures for January were some 2·5°C lower than in the first part of the twentieth century.

Although many parts of the northern hemisphere experienced this notable climatic trend, the southern hemisphere does not appear to have suffered so much. It was not until after 1800 that climatic recession set in, lasting to 1900 or later, and leading to great advances of glaciers in the Andes and South Atlantic islands.

Recent Variations

Reliable instrumental records of climate have been available for about 125 years. The beginnings of these observations in the middle of the nineteenth century witnessed signs of an amelioration of climate after the Little Ice Age. In fact, the warming trend may have begun as early as the 1820s. This improve-

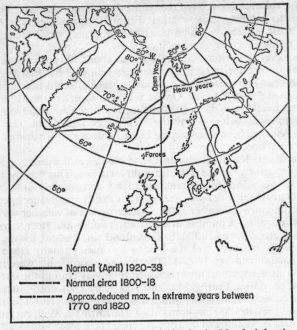

Fig. 17.1. Former limits of pack-ice in the North Atlantic.

ment continued throughout the rest of the century—apart from the decade 1880–90—and into the present century until the 1930s. This period of a hundred years or so saw a rapid retreat of most of the world's glaciers, with a corresponding altitudinal rise in snowline and treeline. Tundra margins and permafrost limits also retreated northwards, and Arctic ports remained ice-free. In other parts of the world, the amelioration was marked by a significant decrease in rainfall, particularly in tropical areas and in south-east Australia.

Table 17.2. Recorded Occasions per Century of Thames Freezing over in London (after Lamb)

Century	Freezings
10th	1
11th	1
12th	2
13th	2
14th	0
15th	1
16th	4
17th	8
18th	6

Unfortunately, as far as mid- and high-latitude countries are concerned, the latest evidence suggests that the warm period of the 1920s and 1930s has come to an end. Since 1940, a slight cooling has taken place, especially in the sub-arctic zone, where the drop in winter temperatures has been of the order of 2 to 3°C. Elsewhere, although glaciers have not started readvancing, the rate of retreat has dropped markedly. A detailed analysis by Lamb of airflow types affecting the British Isles indicates that there has been a decline in the frequency of days with westerly airflow from 38 per cent in 1898–1935, to 30 per cent in 1938–61 (Fig. 17.2). This decline is linked with an increase in northerly airflow, giving more frequent snowfalls. The trend has also been accompanied by a southward shift of depression tracks which has produced a number of cool wet summers in Britain, notably 1954, 1956, 1958 and 1960. Whether this is part of a new major downward trend in climate or merely a small wobble in the amelioration of the last 125 years is something we must discuss in the next section.

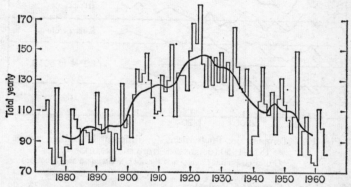

Fig. 17.2. Number of days classified as of westerly type over the British Isles (the superimposed curve shows the course of the ten-year running mean).

Causes of Climatic Change

Regional and Global Variations

To gain some insight into the causes of climatic variation, it is worth noting at the outset that some changes appear to occur on a global scale, but many others, particularly those of a short-term nature, are regional—that is, they are climatic changes characterised by an excess of heat or precipitation in one region, but often matched by a corresponding deficiency in another, without mean global values being affected. As long as the amount of radiation from the sun, the solar constant, really does remain constant, then the mean global values of temperature, evaporation and total precipitation should remain unchanged, unless the composition of the atmosphere (e.g. ozone, carbon dioxide content) is altered in some way.

However, one of the most significant discoveries of the last decade is that these global values are not constant. Many of the European trends we have noted earlier have also affected the rest of the world. For example, the notable warm trend up to 1940 has been world-wide, expressed by a general rise in the mean temperature of the world's oceans of 0·7°C in 60 years. This may seem minor, but is substantial considering the volume of water involved, and has had the side-effect of slightly raising sea-level. Fig. 17.3 shows temperature fluctuations for several latitudinal zones of the world; the general parallelism of the trends is clear.

Regional changes are basically the result of persistent anomalies and these are superimposed on the less spectacular but perhaps more important global changes. They both have a common connection in the behaviour of the general circulation.

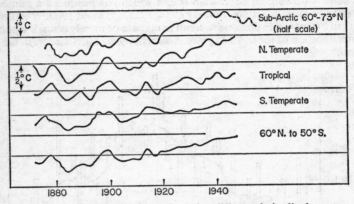

Fig. 17.3. Temperature fluctuations for different latitudinal zones, 1875–1950 (departures from the mean).
(From H. Fløhn, *Climate and Weather*, Weidenfeld and Nicolson)

The Link with the General Circulation

There seems little doubt that the immediate cause of recent climatic fluctuations is linked to the strength of the general circulation, particularly in the northern hemisphere westerlies and in the trade winds. The effect of an intensified atmospheric circulation is to increase the extent of oceanic influence, especially in winter, thereby raising mean temperatures. H. H. Lamb has made a particular study of this thesis. He has shown that the beginnings of climatic amelioration in Europe in the 1820s were linked to a pronounced increase in the vigour of the westerlies over the North Atlantic. This was accompanied by a northward shift in depression tracks, which reached their most northerly mean positions in the 1920s and 1930s. At the same time, the mean pressure of the Icelandic low deepened, whereas that of the Azores high and the winter Siberian high increased, resulting in increased pressure gradients over the North Atlantic and Europe.

In other parts of the world, similar relationships between climate and circulation intensity have been noted. It has been suggested that the increased precipitation in Antarctica may relate to the higher incidence of storms in this century.

In the last 35 years, the atmospheric circulation has been weakening. This first became evident a little before 1940, but it was not until the 1950s and

1960s that a resultant increase in the extent of polar ice in the Icelandic and European sectors became apparent.

Solar Cycles

If we take the argument one stage further and look for the key to these circulation changes, we inevitably return to the fundamental factors of the Earth's energy budget. The link between the distribution of energy and the circulation has been stressed several times in this book. Climatologists have searched repeatedly for periodic trends in the record of climatic fluctuations, and have particularly explored the possibility of a link with the well-known solar cycles. It has to be said that even with the aid of statistical techniques and modern computers, this has met with only limited success. In fact, apart from the daily and annual variations hardly any statistically significant periodicity can be said to exist.

Many attempts have been made to link the well-known **sunspot cycle** of 11·2 years with meteorological events, but results have been conflicting. Typical is the amount of precipitation in East Africa, as reflected in the levels of Lake Victoria, which in the early years of this century appeared to agree very closely with the sunspot cycle. But latterly there have been marked departures (Fig. 17.4). Another possibility is that a significant **double cycle** of 22 years exists within the pattern of the general circulation, and involves, among other things, the frequency of blocking anticyclones over Europe.

If a connection between short-term climatic change and solar output does exist, the mechanics of the connection may be complicated and therefore prone to upset from other factors. Changes in the upper atmosphere may be crucial: a recent hypothesis suggests that ozone becomes more abundant at a certain time in the sunspot cycle. The effect of the increase is to warm the stratosphere and weaken the sub-tropical high-pressure belt and in turn the westerlies circulation, causing a period of lower rainfall.

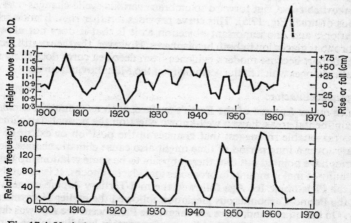

Fig. 17.4. Variations in the level of Lake Victoria (*top*) and the relative frequency of sunspots 1900–1965 (*bottom*).
(From H. Fløhn, *Climate and Weather*, Weidenfeld and Nicolson)

Changes in Atmospheric Composition

Another possible cause of interference in the Earth's atmospheric budget are changes in the composition of the atmosphere. A link between ozone amounts and the sunspot cycle has just been commented upon. Independent of any solar activity, the amount of carbon dioxide seems to bear a relationship with temperatures. Since the beginning of the century, measurements of carbon dioxide have shown an increase of 10 per cent; this would inevitably lead to an increase in heat absorption, and has been cited as a cause of the warm trend since 1900. On the other hand, it has also been suggested that the presence of large amounts of volcanic dust would reflect the sun's radiation and cause a drop in temperatures. The well-known explosion of Krakatoa in 1883 spread a veil of microscopic dust which eventually covered much of the globe.

Astronomical Causes

Attempts have been made to relate climatic change to variations in the Earth's attitude as a planet. This connection has been invoked in a number of theories explaining the glacial/interglacial fluctuations within the Pleistocene Ice Age. The Earth revolves around the sun in an elliptical orbit, and at the same time rotates itself every 24 hours on an axis inclined at $23\frac{1}{2}°$ to the plane of the orbit. At the present time, summer occurs in the northern hemisphere when the Earth is farthest away from the sun (**aphelion**), and the southern summer when it is nearest (**perihelion**). It is known that over long periods of time the shape of the elliptical orbit changes because of different arrangements of planets in the solar system; that the angle of tilt can vary from $21\frac{1}{2}$ to $24\frac{1}{2}°$; and that the seasons will gradually swop over. The latter effect is known as the **precession of the equinoxes.**

It has been argued that any one or all three of these effects could cause considerable variations in the amount of radiation received from the sun, and hence trigger off glaciations. All three effects have been combined in the **Milankovitch curve**, put forward to embrace various cyclic changes over long periods of time (Fig. 17.5). This curve provides a rather rigid framework for glaciations, and one important objection to it is that it does not allow for synchronous glaciation in both hemispheres. However, the theory still receives some support because modern evidence from deep sea cores does show some climatic changes which fit the periodicity of the Milankovitch curve.

Geographical Factors

Since major features of the Earth's surface, such as mountain chains and the continental areas, have a well-known effect on the present-day climate, it seems reasonable to suggest that changes in the position or extent of these features over a long period of time might also cause climatic change. Several theories have pointed out that there appears to be some relationship between the timing of major mountain orogenies and glacial epochs. It is certainly true that the Pleistocene Ice Age followed the mid-Tertiary Alpine orogeny, and that the Permo-Carboniferous glaciation followed the earlier Hercynian orogeny. On the other hand, there are snags: the Pleistocene Ice Age was delayed 25 million years or so after the climax of mountain building in the Alps, although there is some evidence of limited Miocene glaciation elsewhere.

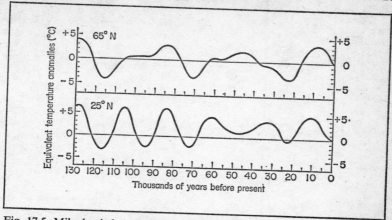

Fig. 17.5. Milankovitch radiation curve: equivalent temperature anomalies for the summer half of the year at 65° and 25° N.

Another suggestion is that glaciation is linked to continental drifting. The Ewing–Donn theory proposes that Pleistocene glaciation was initiated when, relatively speaking, the North Pole reached its present position in the middle of the Arctic ocean, and Antarctica became coincident with the south polar region.

Conclusions

Some tentative progress has been made in recent years in linking short-term regional climatic change with changes in the general circulation of the globe. However, there is no general agreement as to the causes of long-term climatic change. It may be that a combination of causes is the answer. R. F. Flint has suggested a **solar-topographic model**, which puts forward the idea that radiation and relief factors combined are necessary to cause an ice age. But to test this model and all the other hypotheses, we require accurate dating of Pleistocene events, and this reality is not with us yet. However, if we take what we know of the climatic record as the best guide to what may happen in the future, the most reasonable view suggests that we are in the middle of an interglacial, and that another major glaciation is likely, although not for some thousands of years yet.

Suggested Further Reading

Beckinsale, R. P., 'Climatic change; a critique of modern theories', in *Essays in Geography for Austin Miller*, edited by Whittow and Wood, University of Reading, Reading, 1965.

Brooks, C. E. P., *Climate through the Ages* (2nd edn.), Dover Books, New York, 1970.

Lamb, H. H., *The Changing Climate*, Methuen, London, 1966.

Lamb, H. H., 'Climatic variation and our environment today and in the coming years', *Weather*, 1970, **25**, 447–455.

PART THREE
SOILS, PLANTS AND ANIMALS

CHAPTER EIGHTEEN
SOILS

Soils constitute a major element in the natural environment, linking climate and vegetation, and they have a profound effect on man's activities through their relative fertility. Although they are often dismissed as being rather lifeless and merely the static medium for plant growth, soils are very much dynamic entities in which physical, chemical and biological activities are continually taking place.

The scientific study of soils is known as **pedology**; the process of soil formation is referred to as **pedogenesis**. There is no generally agreed statement as to what constitutes soil, but one definition covering most situations is that soil is the upper weathered layer of the Earth's crust affected by plants and animals. A vertical section through this zone constitutes a **soil profile**; in each soil profile there are usually several distinguishable layers or **horizons**, which enable different types of soil to be recognised.

Physical and Chemical Characteristics of Soils

Soil contains matter in all three states: solid, liquid and gaseous. The solid portion is partly organic and partly inorganic. The inorganic, or mineral, part of the soil is made up of particles derived from the parent material, the rocks which weather to form the soil. The organic portion consists of living and decayed plant and animal materials such as roots and worms. The end-product of decay is **humus**, black amorphous organic matter. **Soil water** is a dilute but complex chemical solution derived from direct precipitation and from run-off, seepage, and groundwater. The **soil atmosphere** fills the pore spaces of the soil when these are not occupied by water.

Variations in the composition of the atmosphere give rise to different soil climates, which influence ecosystem functioning (Chapter Nineteen). Soil atmosphere and water are present in inverse proportion to each other. After heavy rain, the ground will initially be waterlogged, with the pore spaces entirely filled with water. But the water moves out rapidly by gravitational movement until the coarser pores are empty and water is no longer supplied to gullies and field drains. The soil is then said to be at **field capacity**. Further removal of water from the soil may occur by evapotranspiration (see page 243), until the pore spaces are largely air-filled and the soil becomes parched.

Soil Texture and Structure

The **texture** of a soil refers to the sizes of the solid particles composing the soil. The sizes range from gravel to clay (Table 18.1). The proportions of the different sizes present vary from soil to soil and from layer to layer. Standard soil textural classes can be defined according to the ratio of sand, silt and clay, and can be represented on a triangular diagram (Fig. 18.1). The corners of the

209

Table 18.1. Soil Texture Grades (in mm)

The International Scale

Clay	Silt	Sand		Gravel
		Fine	Coarse	

0·002   0·02   0·2   2

US Department of Agriculture

Clay	Silt	Sand	Gravel

0·002   0·05   2

triangle represent 100 per cent of each of the three grades. Any point within the diagram defines the percentage proportions of the three grades. For example, near the centre of the diagram, a loam is a soil in which no one grade dominates.

Texture largely determines the water-retention properties of the soil. In a sandy soil, pore spaces are large and water drains rapidly: in a clay soil, the individual pore spaces are too small for adequate drainage. Generally speaking, loam textures are best for plant growth.

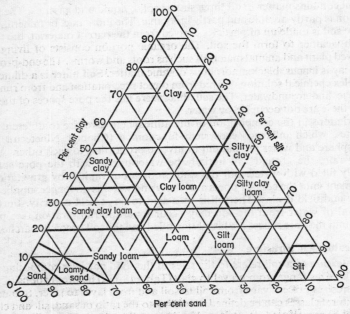

Fig. 18.1. Soil textural classes.

Because of the cementing action of ions in the soil, individual particles in a soil tend to aggregate together in lumps or **peds**. According to the shape of the peds, soils can be described as having a blocky, platy, crumb, or prismatic **structure** (Fig. 18.2). Clay soils tend to have a prismatic structure, whereas at the other extreme some sandy horizons may lack aggregations altogether, and have what is known as a **single grain** structure. The presence of humus helps the formation of a crumb structure.

The soil structure has an important bearing on its ease of cultivation. Soils with a crumb structure are best for seed germination and are said to have a good **tilth**. We attempt to improve soil structure in the garden by forking and raking and on a larger scale by ploughing and harrowing.

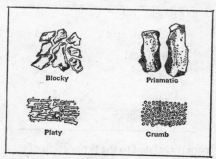

Fig. 18.2. Soil structure types.

Soil Colloids, Bases and Acidity

Included in the clay fraction of the soil are soil colloids—tiny particles with unusual chemical properties. The colloids may be organic, made up of a very finely divided humus, or mineral, in which case they are referred to as clay minerals. Together, the two types make up a **clay-humus complex**. Most soils have more clay minerals than organic colloids. The clay minerals are minute thin flakes but they are of great importance because they are in a state of continuous chemical change, which is fundamental to soil formation.

Clay minerals have a complicated atomic structure and a vast surface area in relation to their weight. Overall, they are negatively charged. This is normally neutralised by the attraction to their surfaces of positively-charged ions (cations) of calcium, magnesium, potassium and sodium. These are known in soil science as **bases**. They are only held loosely in an exchangeable position by the clay minerals, and may be given up in the process of **base exchange** to plants which require them for growth. Some bases are more readily given up than others. In particular, the metallic cations, such as potassium and sodium, tend to be replaced by hydrogen ions. Over a period of time, this makes the soil more acid, unless the bases are replenished in some way. Under natural conditions, the bases are recycled to the soil by the decomposition of plants and animals. Where the vegetation is removed by man by cutting or cropping, the bases can only be fully replenished if supplied artificially in the form of fertiliser. Soft calcareous rocks are often naturally fertile because the rate of weathering of the calcium in the parent material is sufficient to replace the loss of leaching of exchangeable calcium. Lime helps to preserve structural stability in soils.

The bases in the soil are essential as **nutrients** for plant growth. Some nutrients—such as carbon, hydrogen, oxygen and calcium—are required in relatively large quantities; others—such as iron, copper, sodium and magnesium—are only needed in traces but are none the less equally important. Plants also obtain some of their essential elements from the atmosphere. In turn, animals derive their elements through the plants.

Table 18.2.   Soil Acidity

	pH value
Strongly acid	<4·5
Moderately acid	4·5–5·5
Slightly acid	5·5–6·5
Neutral	6·5–7·5
Alkaline	7·5–8·5
Strongly alkaline	>8·5

Soil **acidity** is a property related to the proportion of exchangeable hydrogen in the soil in relation to other elements. The degree of acidity is measured on the logarithmic pH scale (Table 18.2) which ranges from 0 (extreme acidity) to 14 (extreme alkalinity). Few soils reach these limits; most British soils have pH values of between 3·5 and 8·5. A pH value of about 6·5 is normally regarded as the most favourable for plant growth.

Soil Colour

Colour varies considerably in soils and can tell us much about how a soil is formed and what it is made of. The colour of a soil is determined with the aid of a Munsell colour chart, an internationally accepted system of colour designation. In recently formed soils, the colour will largely reflect that of the parent material, but in many other cases, the colour is independent of what lies beneath.

Soils can range from white to black, usually depending on the amount of humus. In cool humid areas, most soils contain a relatively high humus content and are generally black or dark brown, whereas in desert or semi-desert areas, little humus is present and soils are light brown or grey. Reddish colours in soils are associated with the presence of ferric compounds, particularly sesquioxide of iron (Fe_2O_3), and usually indicate that the soil is well drained, although locally the colour may be derived from a red-coloured parent material. In humid climates, greyish or bluish colours reflect the presence of reduced iron compounds, such as FeO, and indicate poor drainage conditions.

Soil Horizons

The recognition of different soil horizons in the soil profile is based on the physical and chemical characteristics of soils so far described. The horizons are labelled according to a system of capital letters with subscript numerals. Although there are several systems in use, most agree on the naming of the major horizons. The system used by the Soil Survey of England and Wales is given in Table 18.3. Originally three parts to a soil profile were recognised,

Table 18.3. Soil Horizon Nomenclature

Current use		Former labels
O (or A_o)	L / F / H	A_{oo} / A_o
A	A_p	A_1
E (or A_e)	E_a / E_b	A_2
B	B_h, B_{1r} / $B_f. B_t$ / B_{fh} / B_s / B_s	B_1 / B_2 / B_3
C		C Parent Material

(Solum spans horizons O through B)

O: Organic Horizon—
 L: Undecomposed litter
 F: Partly decomposed litter
 H: Black, jelly-like humus

A: Mineral-organic layer—
 A_p: Ploughed surface horizon

E: Eluviated horizon—
 E_a: Leached, acid, ash-coloured
 E_b: eluviated basic or slightly acidic

B: Altered or illuvial horizon—
 B_h: humus enriched
 B_f: iron enriched. B_{1r}: iron-pan
 B_{fh}: iron and humus enriched
 B_s: sesquioxides of iron and aluminium
 B_t: illuviated clay

C: Lowest, little altered horizon
A/C, B/C: transitional horizons
A_g, B_g, C_g: gleyed horizons
B_{ca}, C_{ca}: calcium cemented

designated from top to base: A, B and C. Horizons O (Organic) and E (Eluviated) have been separately distinguished more recently. Horizons A, E and B represent the true mineral soil or **solum**; horizon C is the weathered parent bedrock. The processes which produce this arrangement can now be discussed.

Soil-forming Processes

The processes contributing to soil formation involve gains and losses of material to the profile, movement of matter from one part of the profile to another and chemical transformations within individual horizons. In this respect soils can be regarded as another example of open-system phenomena; in theory, if all gains and losses were balanced, the soil will remain in a steady-state equilibrium and be unaltered through time. However, in pedogenesis, not all changes may be towards equilibrium, but may be progressive, and in some cases, irreversible.

The principal soil processes include weathering, translocation, organic changes, gleying, podzolisation and latosolisation. Weathering was considered in Chapter Three; the more important of the other processes will be outlined here.

Translocation

This is not a single process; the term embraces several kinds of movement of material within the soil body, principally by the agent of water. The direction of movement will vary, but may be expected to be predominantly downward in humid environments. The downward movement of material in solution or colloidal suspension is generally referred to as **leaching**; the physical downward washing of clay and other fine particles is known as the **eluviation** of material. Surface layers which have lost material in this fashion are hence called **eluvial** horizons. Eluviation and leaching may move material right out of the soil system, but commonly the solutes and particles are redeposited in the lower parts of the soil profile, creating **illuvial** or enriched horizons.

In arid or semi-arid environments, where potential evapotranspiration exceeds precipitation, movement of soil solution is likely to be upward, drawn by capillary attraction towards the drying surface. The ineffectiveness of leaching ensures that calcium and other solutes remain in the soil. The concentration of calcium in this way, sometimes in layers, is known as **calcification**. In grasslands, calcification is enhanced by the fact that grass uses calcium, drawing it up from the lower soil layers and returning it to the soil when the grass dies.

In extreme cases, where evaporation is very intense, calcium or sodium salts may form a whitish crust at the soil surface, harmful to plant growth. Excessive sodium accumulations are usually the result of the capillary rise of water from a water table that is saline and close to the surface. This process of **salinisation** or **alkalisation** has occurred in some irrigation schemes—for example, in the Punjab.

Organic Changes

Organic accumulation in the soil profile takes place mainly at the ground surface with the decay of plant material. This is gradually broken down or **degraded** by the action of fungi, algae, small insects and worms, reducing the surface litter to its skeletal material. The matter is also decomposed or **humified** into a dark amorphous mass. Under conditions of extreme wetness, humification leads to the formation of **peat**. Over a long period of time, humus itself decomposes in a process known as **mineralisation**, which releases nitrogenous compounds. Degradation, humification and mineralisation are not separable processes and always accompany each other.

Podzolisation (Cheluviation)

This is a process that is widespread in soils that are on the acidic side of neutral. In these cases, because of the different solubility of the various minerals, a situation develops in which the upper horizons of the soil become rich in silica, tending towards pure quartz sand, and take on a characteristic ash-grey appearance. The lower illuvial horizons become rich in sesquioxides, particularly of iron. In some cases the enrichment is concentrated to form an **iron-pan,** a thin but tough horizon of iron oxides. The basic cause of these translocations lies in the leaching action of certain humic acids known as **chelating** agents. These agents are richest in heath plants and conifer needles, which liberate the acids during decomposition, and poorest in grass and deciduous trees growing in base-rich conditions.

Although podzol profiles are usually associated with coniferous and heath-land vegetation, they are found in a variety of other acidic situations throughout the world, except in the tundra.

Gleying

In very wet or waterlogged soils, **anaerobic** (oxygen-deficient) conditions favour the proliferation of specialised bacteria which use up organic matter. They reduce ferric iron to a soluble ferrous state in a process known as **reduction.** This results in the creation of a **gley** horizon, a thick, compact layer of sticky structureless clay. It is also characterised by a bluish-grey or bleached look. The gley horizon normally occurs within the zone of permanent ground-water saturation. Above this, where the soil periodically dries out, the ferrous solutes may be oxidised to form ferric iron. Since this process is not uniform, it gives the soil a mottled or blotchy look, typified by patchy red colours.

Latosolisation

In some respects this process is the tropical relative of podzolisation. Under sustained warm conditions, bacterial action oxidises dead vegetation so rapidly that the chelating action in podzolisation is rare. Instead, in the absence of humic acids, the sesquioxides of iron remain insoluble and accumulate in the soils as red clays. **Lateritic** horizons form where ground-water movements within the soil concentrate the iron and aluminium oxides into a layer just above the water-table. Crusts of indurated laterite at or near the surface are generally relic features formed in a previous episode of soil formation.

Factors Controlling Soil Formation

Many years ago, the Russian pedologist Dukuchaiev firmly established that there were five main factors which controlled the operation of the soil processes just outlined: parent material, climate, organisms, topography and time. Any given soil can be regarded as the product of the interaction of all these factors, although locally one factor may exert a particularly strong influence.

Parent Material

It is a popular misconception that the type of parent material alone determines the kind of soil present. In many soil profiles, only in the C horizon is the soil material similar to the parent material, and major soil types may transcend geological boundaries. The main ways in which parent material is liable to have a lasting effect on soils is through texture and fertility. Thus sandstones and gritstones result in free-draining coarse-textured soils, whereas shales are likely to give rise to much finer soils. Soils vary in their fertility because the parent material supplies many of the initial bases and nutrients in the soil, although they may become redistributed by the soil processes. In particular, the division between calcareous and non-calcareous parent materials has a lasting effect on the soil. Calcareous rocks tend to base-rich soils because of calcification, whereas soils on non-calcareous rocks are liable to podzolisation and a degree of acidity.

Climate

This factor has a major influence in governing the rate and type of soil formation, particularly through precipitation and temperature regime. The main effect of temperature on soils is to influence the rate of chemical and biological reactions. In cool climates, bacterial action is relatively slow and a thick layer of decomposing vegetation covers the ground. In the tropics, bacteria thrive, and although the leaf fall in tropical forests is great, much of this is consumed and translocated down the soil profile.

Absolute precipitation figures have little meaning in soil study, and need to be taken in conjunction with the amount of moisture lost through evapotranspiration. For example, the intensity of the leaching process very much depends on the extent to which precipitation exceeds evapotranspiration. In climates where this is the case, leached soils are sometimes referred to as **pedalfers**, so called because the removal of free calcium leaves behind aluminium and iron oxides. In contrast, where precipitation is less than the potential evapotranspiration—that is, there is a soil-water deficit—the excess of calcium carbonate and other salts gives rise to **pedocals**, soils typical of arid and semi-arid climates. The distinction gives rise to a twofold division of soils in the United States (Fig. 18.3). On a more local scale, differences in precipitation between hilltops and valleys can give rise to different soil types.

The Biotic Factor

The organisms affecting soil development range from microscopic bacteria to large mammals, including man. We have seen that organic matter itself is a basic component of soil profiles, and that humic acids are essential in podzolisation. Besides providing much of the humus, vegetation influences the soil in several other ways. By intercepting direct rainfall and binding the soil with roots, plants check soil erosion. They counteract percolation by transpiration, reducing the effectiveness of the rainfall. Most important of all, plants help to maintain the fertility of soil by bringing bases such as calcium, magnesium and potassium from the lower layers of the soil into stems and leaves, and then releasing them into the upper soil horizons. Different types of vegetation require different proportions of basic nutrients; trees, especially conifers, use little calcium and magnesium, whereas grasses recycle abundant quantities of these. Hence, with these relationships, certain major soil types have specific vegetation associated with them, and a change of vegetation may cause a change in soil.

The influence of animals on soils is both mechanical and chemical. For

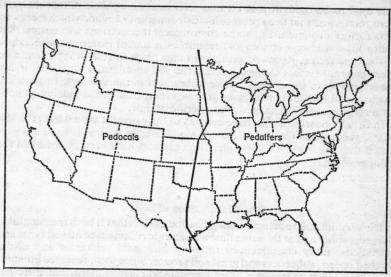

Fig. 18.3. Major soil classes of the United States.

example, earthworms rework the soil by burrowing, and also change its texture and chemical composition by passing it through their digestive systems. Equally, soil characteristics closely determine the type of animal present in the soil; the close relationship between soil systems and ecosystems is further discussed in Chapter Twenty.

Topography

Of the relief factors affecting soils, probably the most significant is that of slope angle. The susceptibility of soil to erosion increases with gradient, and soils on steep slopes are normally thinner than those on flat sites. But situation is also important: a flat hilltop may be a material-exporting site, whereas an equally flat area in a lowland situation will receive downwashed material. In addition, topography affects drainage conditions: soils on hillsides tend to be much better drained than those in valleys, where gleying may take place. Another influence of topography is that it controls exposure to the sun; in middle latitudes south-facing slopes may have slightly different soil conditions from north-facing slopes. The combined result of these topographic effects sometimes gives rise to a soil **catena**, a characteristic gradation of soils occurring within each valley in an area of similar relief.

The Time Factor

It is difficult to be precise about the role of time in soil formation, since soils vary greatly in their rates of development. On porous materials such as sandstones, soil formation is much more rapid than on impermeable materials, at least initially. On glacial tills, a few hundred years may be enough to form a soil; on dense basalt very much longer is likely to be required.

It used to be thought that all thin soils resembling their parent material were young, and that deep profiles indicated maturity beyond which there was little further evolution. But some environmental conditions will ensure that their soils will always be thin and resemble the parent material, independent of the time factor. Renewed evolution takes place in soils when climate or other external factors change, causing the soil to adjust. Frequently, many characteristics of the old soil profile will remain in the new soil. In practice, most soils in mid-latitude regions are **polycyclic**, that is, the balance of soil-forming processes has changed over a period of time.

In Britain, nearly all soils have evolved to their present state within the last 10,000 years, since the end of the last glaciation. In Africa, it has been suggested that soils on the ancient planation surfaces have been undisturbed for millions of years.

The Classification of Soils

It is very difficult to achieve a classification of soils that is both meaningful to the geographer and at the same time an accurate reflection of all soil types and gradations. Early classifications followed biological principles and distinguished orders, suborders and great soil groups. Since then, repeated attempts have been made to produce other hierarchical classifications, but there has been little agreement as to what factors should be used as a basis. Two main types of classification used today may be recognised: those based on the assumed origins of the soil; and those based on the observable morphology of the profile. Examples of each are given below.

The Zonal System

One of the most popular classifications of soils has been the zonal system. This was proposed many years ago by Russian pedologists who recognised the strong relationship between climate, vegetation and soil zones throughout the world. Three main classes of soil are recognised. **Zonal** soils are those that are well developed and reflect the influence of climate as the major soil-forming factor. **Intrazonal** types are well-developed soils formed where some local factor is dominant. **Azonal** soils are those that are immature or poorly developed.

A number of criticisms have been levelled against the zonal concept. One is that the zonal soil type of one climate may well be found in another. For example, podzols, normally recognised as the zonal soil type of cool continental climates, also occur in maritime areas and in the tropics. Another difficulty concerns the azonal class: azonal soils are not necessarily a reflection of the lack of time for development, but may be a result of local factors that have arrested soil development over a long period. A third point is that soil profiles do not always reflect the prevailing climate, and may have characteristics inherited from previous climates.

However, providing its limitations are appreciated, the zonal system is one of the simpler and more intelligible frameworks for studying soil types. It is particularly suitable for describing soil types on a world scale, and has been used as a basis for reviewing major soil types in the latter part of the chapter (see Fig. 18.4).

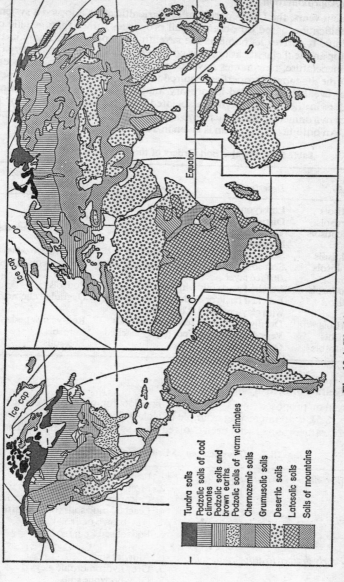

Fig. 18.4. World pattern of soils (zonal types).
(from C. D. Ollier, *Weathering*, Longman)

Tundra soils

Podzolic soils of cool
climates

Podzolic soils and
brown earths

Podzolic soils of warm climates

Chernozemic soils

Grumusolic soils

Desertic soils

Latosolic soils

Soils of mountains

The 7th Approximation

In recent years, the US Department of Agriculture has adopted a system of soil classification based on observed soil properties rather than genetic considerations. It derives its name as the seventh attempt of its authors in their search for an ideal classification. It is controversial because it has a very difficult nomenclature, new names having been coined throughout, and because some of the diagnostic properties required to classify a particular soil type can only be revealed by detailed laboratory analysis. Nevertheless, the primary soil orders in the system are not associated with particular climatic or geographic environments, and it therefore avoids some of the pitfalls of the zonal system. An outline of the system is given in Table 18.4.

Table 18.4. **Major Soil Orders of the 7th Approximation**

Order	Description	Approx. equivalents (Zonal system)
1. Entisols	Embryonic mineral soils	Azonal
2. Vertisols	Disturbed and inverted clay soils	Grumusols
3. Inceptisols	Young soils with weakly developed horizons	Some brown earths
4. Aridisols	Saline and alkaline soils of deserts	Desert seirozems
5. Mollisols	Soft soils with thick organic-rich surface layer	Chernozems, Chestnut, Prairie
6. Spodosols	Leached acid soils with ashy B horizon	Podzols
7. Alfisols	Leached basic or slightly acidic soils with clay-enriched B horizons	Degraded chernozems
8. Ultisols	Deeply weathered, leached acid soils	Red-yellow podzols
9. Oxisols	Very deeply weathered, highly leached	Latosols
10. Histosols	Organic soils	Bog soils

Classification of British Soils

A system proposed by B. W. Avery in 1956 for classifying British soils (Table 18.5), is an interesting example of a classification built up for a specific purpose, rather than having been evolved from some grand concept. The main

Table 18.5. **Classification of British Soils by B. W. Avery**

A. *Well-drained terrestrial soils*	1. Raw mineral soils
	2. Montane humus soils
	3. Calcareous soils (including rendzinas)
	4. Leached mull soils (brown earths, grey-brown podzols)
	5. Podzolised (mor) soils
B. *Poorly drained hydromorphic soils*	6. Alluvial soils
	7. Grey hydromorphic gleys
	8. Gley-podzolic soils
	9. Peaty alkaline soils
	10. Peat bog (including blanket and raised bogs)

emphasis is placed on the overall moisture status of the profile, with the type of humus providing a basis for differentiation. This classification has proved useful in agriculture, since drainage status and humus type are particularly relevant here.

World Pattern of Soils

Zonal Types

Podzols (ash-soil). The effect of the cheluviation process (page 215) is to produce soils with a characteristic bleached E horizon (Fig. 18.5). In some profiles, as in the example in the diagram, the humus is washed down the profile and accumulates as a humus-enriched B horizon, forming a **humus podzol**. In others, there is a marked concentration of iron oxide at this level, forming an **iron podzol**. Sometimes this takes the form of an iron-pan, impeding drainage, and resulting in a **gley podzol**. Podzols of these three types are most wide-

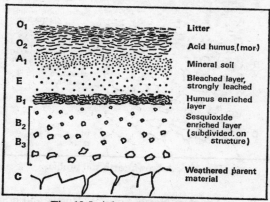

Fig. 18.5. A humus-podzol profile.

spread in the cool climates immediately south of the tundra region, and are found typically in association with coniferous forest, although also under heath.

Brown Earths. These soils are found equatorward of the main podzol zone in milder climates supporting a deciduous forest cover. The soils still exhibit leaching, but of a far less intense nature than podzols. Although free calcium is absent from the upper part of the profile, there is no downward movement of sesquioxides, and their dispersed distribution gives rise to the overall brown colour of the soil. In addition, humus is well distributed throughout the profile and is less acidic than in podzols. Brown earths are widespread in Britain, except in the highland areas (Fig. 18.6).

Tundra Soils. The great variations that exist in the patterns of ground ice in the tundra cause equally complex variations in soils. Where slope conditions are fairly stable, the slow rate of plant decomposition usually results in the presence of a peaty layer at the soil surface. In areas of active slope movement, soils are inevitably thin. In the most extreme conditions where there is no plant growth, the soils are **ahumic**. The brown polar desert soils of the

Antarctic are of this nature. By way of contrast, the birch-forested tundra margins in the northern hemisphere possess Arctic brown forest soils, characterised by a thick dark organic A horizon.

Chernozem, Chestnut and Prairie Soils. The best examples of chernozems and their variants are found in association with steppe or prairie vegetation. The light rainfall of these areas leads to incomplete leaching and the formation of a calcium-rich horizon deep in the profile. Above this is a deep dark layer of soil which can be up to a metre thick. The humus content of this layer is surprisingly often no more than ten per cent, the dark colour being associated with the base-rich mineral matrix. Chernozems have a well-developed crumb structure. The ideal parent material for this soil seems to be loess, which is

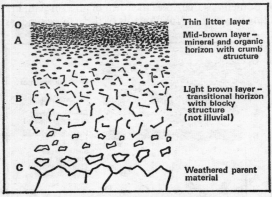

Fig. 18.6. A brown-earth soil profile.

widespread in the mid-west of North America, Russia and northern China. Chestnut soils occur on the arid side of the chernozem belt under a natural vegetation of low grass-steppe. The illuvial carbonate layer is closer to the surface than in chernozems and they have a lower organic content. Prairie soils occupy the transition zone of increasing wetness between chernozems and forest brown earths.

Sierozems of desertic and semi-desertic areas can be regarded as extreme forms of chestnut soils in which lime and gypsum come even nearer to the surface because of upward capillary attraction. Since most of the plants are adapted to arid conditions, there is little leaf fall, and organic matter in these soils is low. However, when irrigated, sierozems can be very fertile, because of their high base status.

Grumusols are dark clayey soils of savanna or grass-covered areas which have a warm climate with wet and dry seasons. There are no eluvial or illuvial horizons, but the whole solum is rich in bases, especially calcium, and hence its dark colour.

Latosols. Soils of intertropical areas are often referred to as lateritic, but strictly speaking, laterite is a weathering product and not a soil type (page 78). Most tropical soils are, however, rich in ferric oxide and are collectively known as latosols. The abundance of sesquioxides of iron and aluminium accounts for the red, brown or occasionally yellow colour of the soil. Profiles are often very deep, but with only poor horizon differentiation. The A horizon makes up the first metre of a typical profile, and is usually mildly acidic with a low humus content. The B horizon commonly extends to fifteen metres or more and is predominantly clayey. Latosolic soils are low in fertility because of the lack of humus and bases.

Intrazonal Types

Hydromorphic soils are those which have undergone gleying and are associated with marshes, swamps or poorly drained upland. Two main types can be recognised, according to the position of the water-table in the profile: **groundwater gleys**, where ground water is below the surface; and **surface-water gleys**.

Calcimorphous soils develop on calcareous parent material. **Rendzinas** are dark, organic rich, and are associated with chalk rock in Britain. Another calcimorphous soil is **terra rossa**, which by contrast is a predominantly mineral soil and is found mainly in the Mediterranean region. The upper horizons are rich in clay and reddish in colour, sharply contrasting with the parent material.

Halomorphic (saline) soils are mostly found in deserts. There are three common types in this group. **Solanchak** (white alkali soils) develop in depressions and exhibit white salt crusts in dry periods. **Solonetz** (black alkali soils) are the product of intense alkalinisation and are characterised by the presence of sodium carbonate. **Solodic** soils develop when leaching in the presence of excess sodium causes the loss of clays and sesquioxides, forming a bleached, eluviated horizon looking rather like a podzol.

Azonal Soils

Immature soils may exist because of the characteristics of the parent material or the nature of the terrain, or simply the lack of time for development. Such situations typically occur in areas where fresh parent material is being deposited or exposed. For example, on active flood-plains, alluvial soils have little or no profile development, because of their frequent burial under new sediments, **Regosols** are composed of dry and loose dune sands or loess. **Lithosols** are accumulations of imperfectly weathered rock fragments on steep slopes where erosion rates remove soil almost as fast as it is formed.

Suggested Further Reading

Bridges, E. M., *World Soils*, Cambridge University Press, London and Cambridge, 1970.
Cruickshank, J. G., *Soil Geography*, David & Charles, Newton Abbot, 1972.
Eyre, S. R., *Vegetation and Soils* (2nd edn.), Arnold, London, 1968.
Fitzpatrick, E. A., *Pedology*, Oliver & Boyd, Edinburgh, 1971.
Jacks, G. V., *Soil*, Nelson, London, 1963.

CHAPTER NINETEEN

THE WORKING OF ECOSYSTEMS

The study of organisms in relation to their environment is known as **ecology**. This science developed in response to the increasing awareness of inter-relationships between plants, animals and their physical habitats. The great voyages of exploration in the late eighteenth and early ninteenth centuries accumulated vast amounts of information about the diversity and distribution of species throughout the world and led naturalists to enquire into the causes of the patterns they found. Early workers acknowledged the importance of the climate in which organisms lived, and attempted climatic classifications of vegetation. One of the earliest of these was by the German biologist Köppen who, in 1918, tried to establish boundaries for vegetational and climatic zones (Chapter Sixteen). At about the same time botanists were studying plant communities and realising the importance of animals in these; their classifi-cations of vegetation included the **index animals** typically associated with certain community types. Similarly zoologists were emphasising the inter-dependence between animals and plants. Early this century naturalists were dividing plants and animals of the world into **biotic associations** or **biomes**. These were large areas in which recognisable associations of species occurred.

Attempts were made to replace this distributional approach with a more functional concept to include plants, animals and their physical habitat work-ing together as a system. A system of organisms functioning together with their non-living environment became known as an **ecosystem**. In this chapter we shall examine the structure and function of ecosystems.

The Structure of Ecosystems

The concept of the ecosystem is very broad and flexible. It can be applied to any situation where organisms function together with their non-living environ-ment in such a way that there is interchange of materials between them even if the system only lasts for a short time. We can look on puddles, fields, forests and the whole world as ecosystems. Ecosystems are **open systems**, and there-fore have flows of energy and materials across their boundaries. Sometimes it is easy to recognise units to study as ecosystems, such as a pond or a wood, but often the boundaries of the system are more arbitrarily placed round the area to be examined, such as a patch of grassland or part of a desert. But what-ever the size of the area to be studied the ecosystem concept is a useful model for examining the structure and function of life.

There are four basic components of an ecosystem. First, the **abiotic** part, which is the non-living environment. Second, the **producers** or **autotrophs**, the green plants capable of producing their own food by using the energy of sun-light to make carbohydrates from water and carbon dioxide; this process is called **photosynthesis**. Third, there are the **consumers** or **heterotrophs**. These are animals which obtain their food by eating plants or other animals. The hetero-trophs in any ecosystem can be divided into groups by their feeding habits:

herbivores eat only living plant material; **detritivores** feed on dead plant and animal material; **carnivores** eat other animals; and **omnivores**, as the name suggests, eat both plant and animal material. Fourth, there are the **decomposers**, such as the bacteria and fungi that promote decay.

An Example of a Simple Aquatic Ecosystem

The four basic components mentioned above can be recognised in very different types of ecosystem and it would be useful at this point to look at one example. Small ponds are a useful starting-point for ecosystem studies because they demonstrate the interrelationships between the abiotic (non-living) and the biotic (living) parts of the system very clearly (Fig. 19.1).

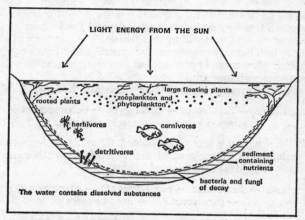

Fig. 19.1. The pond as a simple ecosystem.

Inorganic and organic substances such as water, oxygen, carbon dioxide, calcium and mineral salts form the abiotic part of the system. A small percentage of these will be dissolved in the pond water but most will be present as solids in the sediment at the bottom of the pond. The sediments act as a reserve of nutrients for the plants and animals.

The autotrophs in the pond can be of two main types. They can either be large plants which are rooted or floating, or they can be minute plants, usually algae, called **phytoplankton**. The phytoplankton give the water a green tinge and are distributed throughout the depth of the pond providing there is sufficient light for photosynthesis. They are often very important for producing food in the system and in deep ponds or lakes they usually produce a greater total amount of food than the large plants.

In the heterotroph group of organisms, the herbivores, feeding directly on living plant material, will be of two types, namely the minute **zooplankton**, which feed on the phytoplankton, and the larger animals such as herbivorous fish. The carnivores will include a range of organisms from predaceous insects to game fish which feed on herbivorous fish or each other. There may also be detritivores living at the bottom of the pond, feeding on the dead plant material falling through the water from the autotrophs.

The last main component of the system, the decomposers, will include aquatic bacteria and fungi, and will be distributed throughout the pond. They will be especially prolific at the interface between the water and the sediments where dead plant and animals accumulate. The bodies of these dead organisms will usually decay rapidly due to the continued action of the detritivores and the decomposers.

The Trophic Structure of Ecosystems

The organisation and pattern of feeding in an ecosystem is known as the **trophic structure**. In the example of the pond ecosystem, we have seen that there is a definite arrangement of the main components to form a sequence of levels of eating. This sequence of consumer levels is known as a **food chain**. There are two basic sorts of food chain: a grazing food chain in which the plants are eaten live by herbivores, and a detrital food chain in which the plants are eaten as dead material by detritivores. The two sorts of chain vary in importance in different ecosystems; for example, in a forest ecosystem the detrital chain is often more important, but in a marine ecosystem the grazing chain is usually more important.

Food chains can be simple linear chains which take the form of

plants → herbivores → carnivores → decomposers
(e.g. grass → vole → weasel → bacteria)

but usually the situation is more complicated. Often there are more than four steps in the chain. For example, carnivores could feed on other carnivores so that food chains would take the form of

plants → herbivores → carnivores (1) → carnivores (2)
→ carnivores (3) → decomposers

In addition, some animals consume a wide variety of food; herbivores may eat several sorts of plants, and carnivores may eat several different herbivores and other carnivores. This means that linear food chains will interconnect to form **food webs**. Grazing and detrital food chains often link in this way at the carnivore level. The patterns of food levels in ecosystems can be determined by techniques such as the analysis of the gut contents of animals to see what they have eaten, or by introducing radioactive tracers into plants and monitoring their progress through the heterotrophs.

Fig. 19.2 illustrates the food web associated with the adult herring. This shows that there are many links between the food chains in which the adult herring acts as a carnivore.

Organisms feeding at the same number of steps on a food chain from the autotrophs are said to be at the same **trophic level**. The green plants are at the first trophic level, herbivores the second, carnivores feeding on herbivores the third, and so on. This trophic classification is one of *function*, not of populations, so that an omnivorous species could occupy more than one trophic level.

Energy Flow and the Standing Crop

The energy of sunlight fixed in food production by green plants is passed through the ecosystem by food chains and webs from one trophic level to the

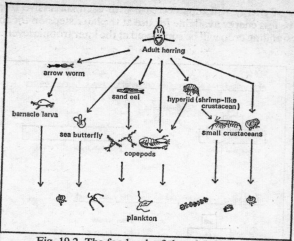

Fig. 19.2. The food web of the adult herring.

next. In this way energy *flows* through the ecosystem. Ecologists have traditionally looked at energy flow in ecosystems in the same way as other scientists have examined energy flow in physical systems. They have applied the first and second laws of thermodynamics. The first of these laws states that energy cannot be created or destroyed. It can only be transformed from one sort to another—for example, light energy into food energy as in photosynthesis. This means that all the energy fixed as food by the green plants must either be passed through the system, be stored in it, or escape from it.

We have already looked at the way in which energy is passed through the system in food chains and webs. Storage of energy in the system is shown by the amount of living material in both the plants and animals present. The amount of living material present is called the **standing crop**. This can be expressed in several ways but is usually shown as **biomass** (living material) per unit area, measured as dry weight, ash weight or calorific value. Ecologists usually look at the standing crop at each trophic level as this is an indication of the pattern of energy flow through the system.

Usually the amount of standing crop in each trophic level decreases with each step on the food chain away from the plants. This can be shown diagrammatically by **trophic pyramids** as in Fig. 19.3. Each bar represents a trophic level and the size of the bar is proportional to the amount of biomass at that level. The characteristic pattern is due to two main reasons. First, the second law of thermodynamics states that no transformation of energy is 100 per cent efficient—there is always some loss of energy as heat. This means that when herbivores eat plants to get food for growth and maintenance of their bodies, they will not be able to use all the food energy in the plant material. In converting plant substances into animal substances there will be loss of energy as heat, which will escape from the system. Therefore there will be large losses of energy *between* trophic levels. Second, there will be energy losses *within* each trophic level. All organisms must respire to live; **respiration** involves the oxidation of carbohydrates to release energy. Therefore energy will be lost from the trophic level by the respiration of the organisms in it.

The flow of energy will thus decrease with each successive trophic level. There will be less energy available for use at the later steps on the food chains and so less standing crop will be supported at the later trophic levels.

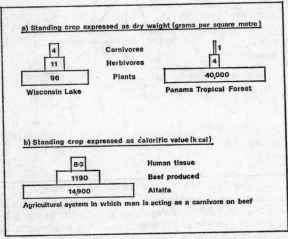

Fig. 19.3. Examples of trophic pyramids.

Productivity

In ecosystems the rate of production of organic matter is known as productivity. **Primary** productivity refers to production at the autotroph level, and **secondary** productivity refers to production at the heterotroph level.

We can divide productivity still further into **gross** and **net**. Gross productivity is the total amount of organic matter produced, and net productivity is the amount of organic matter left after some has been used in respiration. Primary gross productivity will depend on the efficiency of photosynthesis and the amount of light energy coming into the system. The intensity and duration of sunlight varies globally (Chapter Ten) so that the potential for gross primary productivity will vary greatly with different ecosystems. Vegetation at the equator will receive far more light energy in a year than vegetation at high latitudes. The efficiency of photosynthesis itself depends on many factors such as temperature, the availability of nutrients and the age and species of individuals. Net primary productivity will be determined by the relative rates of respiration (using up carbohydrates) and photosynthesis (producing carbohydrates). In reasonable conditions photosynthesis proceeds up to thirty times faster than respiration but it must be remembered that it only occurs in the light.

Very few detailed quantitative studies of primary productivity have been conducted, but those that have all reveal very low rates. Of the light energy impinging on the surface of vegetation as little as one to five per cent may be trapped as food energy. Yet despite these low figures enough energy is trapped to maintain all life. It is interesting that where comparisons have been made, agricultural systems have been found to be far less productive than natural systems.

Secondary productivity will depend on the conversion of plant substances to animal substances. The efficiency of transfer of energy from one trophic level to the next is known as **ecological efficiency**. The efficiency of transfer from autotrophs to heterotrophs is low. Most studies now estimate it to be about ten per cent in natural ecosystems. The majority of animals have their greatest rates of net productivity when they are young as this is the time when they are most vulnerable. It is an ecological advantage to grow quickly in order to compete with other animals for survival. Even at their most productive, animals rarely exceed a 35 per cent efficiency of energy transfer.

At first it may seem strange that productivity in ecosystems should be so low, but it must be remembered that there are many pressures in the environment to influence the evolution of species. Selective forces for breeding, escaping from predators or maintaining territory may take precedence over the importance of ecological efficiency.

Modelling Energy Flows and Productivity

For convenience many ecologists have modelled energy flowpaths by grouping organisms by their trophic levels and indicating the energy flow as relationships between these levels. The levels can be shown diagrammatically as boxes for plants, herbivores, etc., connected by arrows or bars representing energy flow. The bars can be of varying width, proportional to the amount of energy flowing. In reality this is extremely difficult to do as species often have many different roles in the system and cannot be neatly placed in trophic levels. Very few so called 'carnivores' are species that feed only as true carnivores. In nature, energy flows through the ecosystem by a complex of encounters between species, leading to interaction between trophic levels. Energy flow paths are rarely simple. Frequently there are feedback loops; one species may feed on the faeces of another so that energy in the faecal material does not go to the decomposers but is taken back into the system at a lower trophic level.

Many ecologists have modelled energy flow and productivity by using a **hydraulic analogy**. Energy is shown as flowing in pipes of various widths between trophic levels. Loss of energy in conversion or by respiration is shown as bouncing off the trophic level (Fig. 19.4). Although this model gives a neat visual impression of the functioning of an ecosystem it should be remembered that it is an extreme abstraction of reality and as such its use is limited.

Ecological Niche and Species Structure

The role that an organism takes in the ecosystem is known as its **ecological niche**. Species vary in the breadth of the roles performed. Some animals are **specialists** in their feeding habits—such as the koala bear, which eats only eucalyptus leaves—and others are **generalists**, consuming a wide variety of food. Some extreme generalists such as *Euglena*, a unicellular alga, can function either as autotrophs or heterotrophs. Usually the specialists in the system are more efficient at using their resources and therefore became abundant when their particular food supply is ample. Despite this advantage, they are vulnerable if conditions change so that their food becomes rare, as they will not be able to switch to an alternative. The majority of animals occupy a broad ecological niche and so are generalists in their feeding habits. Most exhibit a

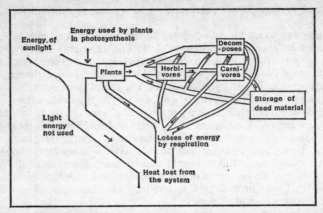

Fig. 19.4. The hydraulic analogy in modelling energy flow. Energy is imagined as being channelled through pipes whose width is proportional to the amount of energy flowing.

scale of preference within their range, switching from one thing to the next if supplies are short. Generalists may never become as locally abundant as specialists but are far less vulnerable to environmental change.

The ecological niche of a species may vary through its distribution in relation to its **habitat**, which is the name given to the place where an organism lives. Factors such as availability of different foods and competition from other species will influence the role of an individual. Man is an example of this, operating as a herbivore, a carnivore and an omnivore in different places.

Similarly, different species may occupy the same niche in equivalent ecosystems. Grazing kangaroos on Australian grasslands and cattle on American grasslands both function as herbivores in ecosystems where the main autotrophs are grasses.

Most ecosystems contain a great variety of ecological niches and so have a mixture of generalists and specialists.

The **species structure** of an ecosystem refers to the numbers of species present, their relative abundance and diversity. Characteristically ecosystems contain a few species that are relatively common, having large numbers in their populations or a large amount of biomass in their standing crop, and a large number of species that are rare. Ecologists often call the few common species the ecological **dominants**, and the rare species the **incidentals**. Neither of these terms is very accurate as the ecological dominants may not actually impose the greatest effect on the rest of the system and the incidentals may be extremely important ecologically.

Species with a high frequency would probably be specialists exploiting prevailing conditions and so becoming abundant. If the environment changes, perhaps becoming wetter or hotter, they may not be so successful due to their narrow ecological range. Some rarer species may be able to exploit the new situation better and so increase in abundance. The incidentals thus form a **reserve of adaptability** in the system which allows a continuation of energy flow and productivity if environmental conditions change.

Nutrient Cycles

The flow of energy in ecosystems is one-way. Light energy is continually being converted into food energy which is stored in the system, passed through it or dissipated as heat. In contrast, the nutrients which are needed to produce organic material are circulated round the system and are re-used several times. All natural elements are capable of being absorbed by plants, usually as gases from the air or as soluble salts from the soil, but only oxygen, carbon, hydrogen and nitrogen are needed in large quantities. These substances are known as **macronutrients** and form the basis of fats, carbohydrates and proteins. Other nutrients, such as magnesium, sulphur and phosphorus are needed in minute amounts and are known as **micronutrients** or **trace elements**. Both macro- and micronutrients are taken into plants and built up into new organic material. The plants may die or be eaten by an animal but in either case the nutrients will be returned to the abiotic environment when the body of the organism is decomposed.

The functioning of all ecosystems depends on the circulation of nutrients. If nutrients were not cycled, available supplies would become exhausted and productivity would be inhibited. Little systematic work was done to determine the pathways of nutrient cycling until the 1930s but now the main routes and mechanisms of cycles have been deciphered.

Nutrient cycles can be presented in the framework of a model in which each cycle has a **reservoir pool**, which is a large, slow-moving non-biological component, and an **exchange pool**, which is a smaller, more active portion where the nutrient is exchanged between biotic and abiotic parts of the ecosystem. There are two basic types of cycle, **gaseous** ones, in which the reservoir pool is the atmosphere, and **sedimentary** ones, in which the reservoir pool is the Earth's crust. All nutrient cycles involve interaction between soil and the atmosphere, and include a wide variety of organisms encouraging movement of the element along food chains to the decomposers. However, cycles do show contrasts in the rates and completeness of exchange of the element between abiotic and biotic parts of the ecosystem. Generally the gaseous cycles are more complete than the sedimentary ones. In Nature, both types of nutrient cycle are assumed to be stable but the sedimentary types are more susceptible to disruption, particularly by human interference.

The Phosphorus Cycle

The phosphorus cycle is an example of a sedimentary cycle which is easily disrupted (Fig. 19.5). Phosphorus is relatively rare in nature but is essential for plant and animal growth. The exchange pool involves cycling between organisms, soil and shallow marine sediments. Phosphates in the soil are taken into plants for protein synthesis and are passed through the food chains of ecosystems. When plant and animal bodies and their excretory products decompose, the phosphorus is released to the soil where it can either be taken back into plants or washed out by rainfall into drainage systems which ultimately take it to the sea. If this happens it will be incorporated in marine sediments and so lost from the exchange pool. One important route for the rapid return of phosphorus from these sediments occurs where there are upwelling ocean currents. These bring phosphorus to the surface waters, where it is taken into marine food chains. Sea-birds feeding on fish in such areas have phosphorus-

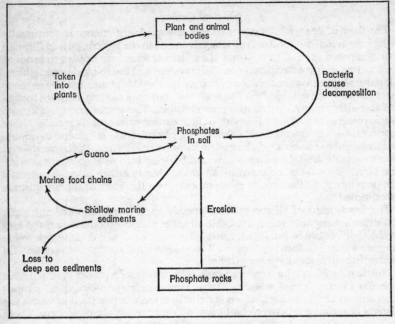

Fig. 19.5. The phosphorus cycle.

rich droppings called **guano** which can be used as a fertiliser. However, the return of phosphorus in this way tends to be localised, occurring for example on islands off the Peruvian coast, and it is estimated that it accounts for less than three per cent of the total amount lost from the land.

The depletion of phosphorus from the exchange pool is compensated very slowly by the release of the element from the phosphate rocks of the reservoir pool. This occurs by the process of erosion and weathering.

The phosphorus cycle can be easily disrupted by the use of phosphate fertilisers in modern agriculture. Most manufactured phosphate fertilisers are produced from phosphate rocks but are rapidly lost from the exchange pool to marine deposits as they are easily leached from the soil. This could lead to serious deficiencies in available phosphorus for agriculture in the future.

The Nitrogen Cycle

The nitrogen cycle (Fig. 19.6) is an example of a gaseous type. It is probably the most complete of the nutrient cycles. The reservoir pool is the atmosphere and the exchange pool operates between organisms and the soil. Nitrates in the soil are absorbed by plants and pass through food chains. Ultimately they are released as ammonia when organic material is decomposed. The ammonia is changed back to nitrates by the action of bacteria. If the nitrates are not reabsorbed by plants they may be lost from the exchange pool in two ways: first, by leaching from the soil to shallow marine sediments (in this case they may be returned in the droppings of marine birds in the same way as phos-

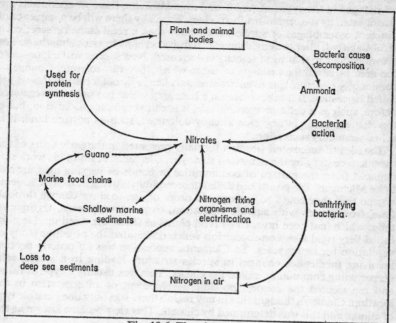

Fig. 19.6. The nitrogen cycle.

phorus); second, nitrates may be lost from the soil by being broken down by **denitrifying** bacteria, and the nitrogen contained in them being released to the atmosphere.

Atmospheric nitrogen in the reservoir pool cannot be used directly by most plants. It has to be made into a chemical compound such as a nitrate before it is available to the exchange pool. The conversion of gaseous nitrogen to nitrate occurs in two main ways. Some can be fixed by electrical action during thunderstorms, but most is converted by nitrogen-fixing organisms. These are mostly bacteria, algae and fungi, and either operate by themselves in the soil or in an association with a plant, particularly those in the legume family, such as clover. Associations between nitrogen-fixing bacteria and legumes form root nodules which produce nitrates and exude some to the soil. Because of this they are extremely important in crop rotations to maintain soil fertility.

Succession and Climax

The nutrient cycling and energy flows in ecosystems will depend on the development of the system and the species present. The species forming a community at any specific place and time will be a selection from those which can tolerate the prevailing environmental conditions. Organisms can modify their environment by influencing soil formation and microclimates so that even without climatic change or disruption the physical conditions in any location will not be constant. Consequently, the species structure of an eco-system will not be constant. This is particularly evident if an ecosystem is developing on a bare land area such as volcanic ash or sand dunes. There will be changes in the types of plants and animals living there as conditions are ameliorated by the organisms themselves. Typically there will be a sequence of different assemblages of species, each known as a **seral stage** or **sere**; each seral stage will alter the environment slightly, perhaps by enriching the soil or growing taller so that wind velocity is decreased. New species will migrate into the area as conditions change to those which they can tolerate. Change of community structure due to environmental change through time in this way is called **succession**. If it takes place on a bare area which has not been vegetated before, such as a delta or volcanic ash, it is called primary succession, but if the changes occur in any area already colonised, such as pasture land, it is called secondary succession.

The idea of succession was first formally presented in the early years of the twentieth century by the American ecologist, Clements. A lot of his work was deduced from the pattern of communities in ponds of varying ages around Lake Michigan. The ponds had different communities living in them. Clements argued that they could be arranged in order of age to show change through time, from ponds with aquatic vegetation, to those with reeds, through to those which had been invaded by land plants as the water dried up. He published very rigid views on succession which dominated the objective study of vegetation for many years. To Clements, succession was an orderly process involving **predictable** changes in species structure leading to a stable, self-perpetuating community, called the **climax**. This was usually a type of forest and represented the greatest possible development of an ecosystem in the location. Clements thought that in any region there was only one possible type of climax and this was determined by climate. This view became known as the **climatic climax** idea.

Clements had great influence on ecology, and his contemporaries were guided by his **deterministic** approach to vegetation change. There was some early opposition to his views on climax, which we will discuss later, but it is only recently that his basic concepts on succession have been reviewed.

Current Ideas on Succession

Succession still provides a useful model to examine the major changes in an ecosystem as it develops towards stability, but it is now recognised that changes in species structure and the sequence of communities are not predetermined. There is a large **random** factor which renders prediction in succession difficult. The migration into an area to form a community may be largely a function of chance events (Chapter Twenty-One). However, there are recognisable trends in succession.

As the ecosystem develops, energy flow will increase in relation to productivity. Productivity itself will increase in the early seral stages but will decrease as a stable community is attained. This can be understood if we examine the structure of the stable community. It will contain the tallest plants of the succession, frequently shrubs or tall trees which are long-lived. They have evolved to have maximum productivity when young so they can compete against other species for survival. When mature they are large and achieve ecological dominance but do not increase their biomass significantly. When communities become stable they are dominated by these large, long-lived plants which have low net productivity when mature, which means that the ecosystem as a whole has low productivity. In the early seral stages, gross productivity will be greater than community respiration so that there will be an increase in biomass. In the later stages total community respiration will increase as there will be more biomass to maintain, so net productivity will be negligible.

One of the most vital changes in ecosystem function through succession is the pattern of nutrient cycling. As succession takes place, more nutrients will be cycled and held in the standing crop. The efficiency of nutrient cycling will increase as succession proceeds towards stability. This will be largely due to the increase in complexity of the system through time. The species number will increase through early stages, leading to the development of food chains and webs. As stability is achieved a few species usually became dominant and eliminate some rarer species by competition. Consequently, there is usually a decrease in species numbers as the ecosystem matures.

The Climax Concept

The mono-climax idea of Clements was widely accepted. He argued that for any major regional zone only one type of vegetation climax could be expected and this would be determined by climate. At any stage in seral development the succession might be halted by arresting factors such as soil conditions to produce a **sub-climax**, but these were assumed to be transient. Many ecologists thought this approach was too rigid. In particular, the famous British plant geographer, Sir A. Tansley, criticised the mono-climax idea as too simple and unworkable. He suggested an alternative **poly-climax** approach, in which climax communities could be determined by a range of factors such as soil, topography and salinity so that a mosaic of climaxes could be present in one climatic area.

More recently workers have challenged the climax ideas as a whole. For example, Whittaker considers that there is no absolute climax determined by environmental conditions. Any community is an expression of its particular habitat and is therefore unique. Communities are constantly adjusting in response to the physical environment but it is doubtful whether they ever attain equilibrium with it. There will be considerable timelags through breeding cycles of the community dominants—most trees take at least 15 years to mature. Climates are constantly oscillating so that readjustments in ecosystem species structure are initiated but by the time these have come into effect the climate may have changed again. Therefore it is doubtful if the concept of a stable climax community has any validity. Many ecologists now think that communities should be viewed as an expression of all the environmental factors, such as climate, topography and soil, which influence the ecosystem. In the next chapter we will examine the influence of these environmental factors in detail.

Suggested Further Reading

Colinvaux, P., *Introduction to Ecology*, John Wiley, New York, 1973.
Odum, E. P., *Ecology*, Holt, Reinhart & Winston, New York, 1966.
Phillipson, J., *Ecological Energetics*, Edward Arnold, London, 1966.
Watts, D., *Principles of Biogeography*, McGraw-Hill, London, 1971.

CHAPTER TWENTY

ENVIRONMENTAL CONTROLS

We have seen in the preceding chapter that the plants and animals forming the biotic part of an ecosystem will be the species which can tolerate the prevailing environmental conditions. No living thing exists in isolation. There are complex interactions and interdependences between organisms and their environments and among organisms themselves. It is very necessary to pay attention to the influence of the environment when examining ecosystems or studying individual species.

Factors which have some effect on the life of an organism at some stage in its development are called **environmental factors**. These can be divided into groups as follows: first, a **climatic** group, which includes conditions of light, temperature, water availability and wind; second, **topographic** influences of slope angle, orientation and altitude; third, **edaphic** (soil) factors, especially pH and fertility; and fourth, **biotic** controls, such as competition between species and the effects of grazing.

These groups are themselves interrelated so that it is extremely difficult to isolate the influence of individual factors. For example, topography and climate will influence soil development; and climate and soil will influence the pattern of biotic controls by determining the species which may inhabit a particular place and compete there for survival.

The Concept of Tolerance

One of the most influential developments in the study of environmental factors took place in 1911 when Shelford presented his concept of tolerance. He put forward his ideas as a series of 'laws', following the scientific fashion of the time. In these laws he stated that plants and animals are limited to the environments they can endure. Every species has a minimum and maximum requirement for each environmental factor, and a range of tolerance between these limits which includes an optimum condition. This can be shown graphically (Fig. 20.1) as a bell-shaped curve. Ranges of tolerance can vary both between species for each factor and within one species for different factors. Narrow ranges of tolerance can be indicated by the prefix steno- (as in stenothermal, meaning a narrow range of tolerance to temperature) and wide ranges of tolerance can be indicated by the prefix eury- (as in eurythermal, meaning a wide range of tolerance to temperature). Therefore a particular species could be eurythermal, stenophagic (narrow range of tolerance to food) and euryhydric (wide range of tolerance to water conditions). Shelford's laws state that the species with wide tolerance ranges for all factors would be the most widely distributed.

It is interesting that these laws also state that reactions to individual factors may be interrelated. If conditions for one factor are not optimum then the tolerance to another factor may be decreased; for example, tolerance to high temperatures may be less if an animal does not have sufficient water.

Shelford also stated in his laws that environmental factors are most limiting in the reproductive stages; for animals this refers both to the reproducing adult as well as the offspring. Perhaps the most startling realisation presented in Shelford's laws is that organisms rarely live at their optimum location. This would be a difficult situation for them to achieve because they may be excluded by competition from other species and also due to the fact that the environments are constantly changing.

Although most ecologists now consider Shelford's approach to be too rigid, the laws are basically correct. The presence and success of an organism does depend on a complex of conditions, but any factor which approaches or exceeds the range of tolerance may be a **limiting factor** in the distribution of the species.

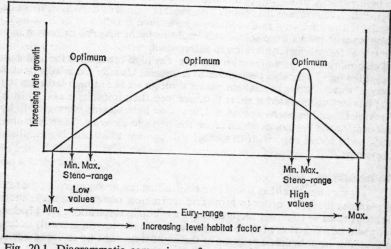

Fig. 20.1. Diagrammatic comparison of ranges of tolerance of organisms to habitat factors.

Shelford's laws of tolerance gave great impetus to an era of 'toleration ecology' in which extensive laboratory experiments were conducted to determine exact limits of tolerance for environmental factors in different species. This work did not solve the basic question of why these sets of tolerances should have evolved. Many ecologists now feel we should examine the way organisms have evolved to take advantage of particular resources and hence by necessity adopt tolerance ranges, rather than concentrate on these tolerance ranges themselves. Even so it cannot be denied that thinking of plants and animals in terms of their habitat ranges is a useful approach for ecosystems analysis. In the rest of this chapter we shall be looking at the ways environmental factors operate to influence ecosystem development.

Light

Light is an extremely important environmental factor because it is the vital source of energy for ecosystems and it can also act as a control of functions such as reproduction and migration. Excess light can be a limiting factor in ecosystem development by damaging plant tissues and decreasing productivity.

The influence of light varies with its three main aspects: its quality (that is, wavelength composition), its intensity and its duration (day length). We shall examine each of these in turn.

The Quality of Light

The amount of energy available for primary productivity will be partly determined by the quality of light. In photosynthesis light energy is absorbed by pigments, the most important of these being green **chlorophyll**, which absorbs red and blue light but reflects green. In terrestrial plant communities and in aquatic ecosystems tall or floating green plants will filter light so that the light passing through to plants beneath will be mainly in the green wavelength band. This means that plants living within woods or deep in water must be adapted to surviving in conditions where there is little red or blue light. Many aquatic plants have solved this problem by having red or brown pigments **(phycoerythrins)** in addition to chlorophyll.

The quality of light varies with altitude. On high mountains the invisible ultraviolet light (very short wavelength) is intense. Usually ultraviolet light is screened out by ozone as sunlight passes through air but at high latitudes the light has only penetrated a short distance into the atmosphere and so little ultraviolet light has been dissipated. Ultraviolet light retards plant growth by deactivating the hormones which cause the stems to elongate. Therefore only species which can cope with this effect will be able to inhabit high-altitude ecosystems.

The Intensity of Light

This is very important as it will be a controlling factor in governing the rate of photosynthesis. In order to grow and reproduce, plants must make more carbohydrates by photosynthesis than are used up in respiration. Net productivity will be a function of these two processes. As photosynthesis only occurs in the light, carbohydrates respired in the night-time must be replaced before there is a net gain in organic matter each day.

The point at which photosynthesis balances the energy used up in respiration is known as the **compensation point** (Fig. 20.2). Plants vary in the light intensity required to achieve their compensation points. Species which grow best in high light intensities are known as sun-plants or **heliophytes**. These have high rates of respiration and photosynthesis and so have high compensation points. In comparison, those which grow best in low light intensities are called shadeplants or **sciophytes**. These have low rates of respiration and photosynthesis and therefore have low compensation points. Plants may change their tolerance to light intensity during their life-span; trees often function as sciophytes when seedlings, and then as heliophytes when they grow to the forest canopy and are exposed to bright light.

There is far more spatial variation in light intensity than in quality. Perhaps the most obvious variation occurs globally with latitude (Chapter Ten); low latitudes receive more intense light than high latitudes, and therefore tropical ecosystems will have a higher energy input than arctic ecosystems. This is an important aspect to remember when examining world vegetation types. On a smaller scale, light intensity varies with topography. The orientation and angle of slopes may have great effect on the local distributions of species. This effect may be an indirect consequence of light intensity because of its influence on other micro-climate factors such as temperature and relative humidity.

Water reflects and absorbs a great deal of light even when clear and still. In aquatic ecosystems light intensity will decrease geometrically as depth increases arithmetically, hence primary productivity will be restricted to a shallow surface layer where plants can achieve their compensation points.

Light intensity also varies within plant communities. Tall plants will shade those growing beneath them so that in forests and woods there will be a range of ecological niches for tolerance to light intensity. This frequently leads to the development of layers of vegetation, typically trees, shrubs, tall herbs and ground plants, each layer filtering out some light and tolerating the intensity regime prevailing.

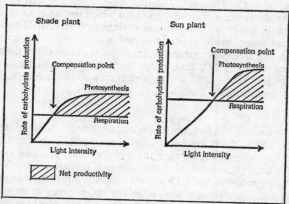

Fig. 20.2. The tolerance between respiration and photosynthesis with varying light intensity.

The Duration of Light

Many aspects of behaviour in plants and animals, such as flowering, migration and mating, are affected by day length. The response of organisms to variation in day length is called **photoperiodism**. The reaction may be linked to other factors such as temperature but the specific stimulus of light is often vital.

Flowering in many species of plant is initiated by a certain number of hours of darkness. Plants can be divided into three latitudinal groups on this basis: first, **long-day** plants (flowering in the long days of temperate summers); second, **short-day** plants (flowering in the short days of the tropics); and third, **day-neutral** plants which have no definite day length requirements. This last group is frequently found in high latitudes where there is continuous light in the summer season. The photoperiodic response is very important in the distribution of species as obviously a long-day plant will not do well in the short days of the tropics and vice versa.

Photoperiodism also occurs in many groups of animals. In birds it frequently stimulates reproduction and migration. Similarly, insects may be influenced in their diurnal activity patterns, their time of metamorphosis and hibernation. Fish respond to a wide range of external stimuli to coordinate seasonal breeding and migration, but for fish living in the light zones near the surface the vital stimulus is frequently day length. Many temperate fish are inactive in the winter and then become photosensitive at the prespawning stage in the spring. At this time day length influences the development of mating colours and reproductive organs.

Temperature

Temperature is a universally important environmental factor both for its direct effects on organisms and for its indirect effects in modifying other factors such as relative humidity and water availability. Each species has its own minimum, maximum and optimum temperatures for life but the actual limits at any time will vary with such things as the age of the individual and water balances in the body. Generally, aquatic plants and animals have narrower tolerance ranges for temperature than those which live on land. This is mainly because there is far more temperature variation in terrestrial ecosystems. In these one can detect broad latitudinal patterns for minimum and maximum temperature requirements; for example, tropical plants often do not tolerate temperatures below 15°C, and most temperate cereals are intolerant of temperatures less than −2°C, whereas evergreen coniferous forests may withstand many degrees below freezing.

The Effect on Plants

All plants are adapted to grow in a relatively limited range of temperatures. Usually the upper limits are most critical although lethal temperatures are seldom reached because plants are cooled by a process of water loss from their leaves called **transpiration**. If the water supply is inadequate wilting will occur and the plant could overheat. Death may then result from a combination of desiccation and high temperatures which cause enzymes to become inactive. Plants living in hot climates frequently have morphological features such as a thick covering or a reflective surface which enable them to avoid heat damage.

If the temperature drops below that required for photosynthesis, plants will cease to grow. In very cold conditions there may be chemical changes in the plant tissues such as the precipitation of proteins but at all costs plants must avoid water freezing in their cells. Ice crystals are very large compared to plant cells and cause irreparable damage by puncturing the delicate cell membranes. There is a variety of strategies plants can adopt to avoid this happening and we shall discuss these later.

Besides direct effects, low temperatures may inhibit plant growth indirectly by inducing a **physiological drought**. This occurs when water in the soil cannot be absorbed by plants either because it is frozen or because the temperature is too low for uptake. The tolerance of plants to low temperatures varies both with the degree and duration of cold. Laboratory experiments have shown that in many cases long periods at a few degrees below freezing may be more damaging than a short time at a much lower temperature.

Plants native to climates with marked seasonal differences in temperature must be capable of surviving adverse conditions in the cold season. Survival depends on the storage of food and protection from desiccation and cold damage. Annual plants do this by completing their life cycle in the summer season and overwintering as a seed. Perennial plants may survive in a state of dormancy through the winter, perhaps losing their aerial parts and overwintering as bulbs, rhizomes or corms. Trees and shrubs may adopt the deciduous habit, losing their delicate leaves in the cold season, or if they are evergreen, cold damage and transpiration losses can be minimised by increasing the density of the cell sap, thus lowering its freezing point. In many cases a certain amount of warmth is necessary to break the winter dormancy. This is known as **vernalisation**.

Temperature has direct effects on plant physiology. For any species the processes of respiration and photosynthesis have different optimum temperatures. Normally the optimum temperature for respiration is higher than the optimum for photosynthesis (Fig. 20.3). Usually photosynthesis proceeds at

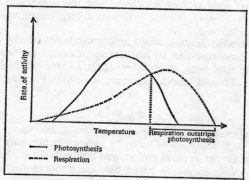

Fig. 20.3. Typical rates of respiration and photosynthesis in relation to temperature.

about thirty times the rate of respiration so that the compensation point is reached and net productivity achieved. However, if the temperature is near the optimum for respiration all the time the rate of carbohydrate consumption will increase and net productivity will be negligible. Because metabolic activities have different optimum temperatures, plants tend to thrive in an environment of varying temperatures, and have depressed growth in constant conditions. The response of plants to rhythmic fluctuations in temperature is known as **thermoperiodism**. The optimum fluctuations for individual species can be determined by laboratory experiments; for example, it has been found that tomato plants grow best with day temperatures at 20°C and nights at 10°C.

The Effects on Animals

The distribution and functioning of animals is also markedly controlled by temperature. Animals can be divided broadly into two groups: the **cold-blooded** ones, which are at the same temperature as their surroundings, and the **warm-blooded** ones, which maintain a constant body temperature by heat regulation. The cold-blooded animals will only be active when the environmental temperatures are suitable. Below this range they may survive in an inactive state but above it their survival is limited. Warm-blooded animals have a distinct advantage as they can remain active despite the temperature of their surroundings. However, they still have limits of endurance and may have to avoid adverse conditions.

The cause of heat death in animals is obscure. Possibly it is due to a combination of increased oxygen requirements as the rate of respiration increases and the coagulation of proteins in the tissues. Similarly the effects of excess cold are not clearly understood. As temperatures decreases the rate of metabolism will decrease and the animals will become inactive. If ice forms in the cells there will be irreparable damage in the same way as in plants.

Both cold- and warm-blooded animals can adjust to extremes of temperature to some extent. In cold-blooded animals this can be accomplished by behavioural patterns; for example, desert lizards move in and out of shade in response to heating and cooling. The warm-blooded animals have a whole range of potential mechanisms, which include the possession of insulating layers of fur or fat and a shivering reaction to keep warm, and sweating or licking their fur to lose heat.

Some warm-blooded animals avoid cold periods entirely by becoming inactive. This can be done either by just sleeping through the winter, as in the case in some species of bear, or by hibernating. In true hibernation the body behaves like that of a cold-blooded animal and follows the temperature of the environment.

Water

Water availability may often restrict ecosystem development because most organisms need large amounts of water to survive. It not only forms a large percentage of the tissues in plant and animal bodies but it is also essential for transport and cooling. In plants, water provides support and is essential for photosynthesis.

The higher plants take in water through their root hairs and transport it up the body in special conducting tissue called **xylem**. Most of the water absorbed passes out of the plant in transpiration from special pores in the leaves known as **stomata** (singular **stoma**). The actual rate of water loss by transpiration will vary with relative humidity, air movement, size of the leaves and the size of the stomata. This means that the water requirements for plants will vary both with environmental conditions and among different species. If we examine the ratio of growth to water transpired, known as the **transpiration efficiency**, we would find that for every one part of dry weight produced, species may transpire between two hundred and a thousand parts of water. This is a useful guide to the water requirements of different species.

If there is insufficient water, plant cells lose their rigidity and the plants may wilt. Stomata close, helping to prevent further water loss. The plant may remain in this condition for a long time without damage, providing the temperature is not excessively high.

Some early ecologists divided plants into groups based on their water requirements and tolerance. One of these, which is still frequently used, was composed by Warming in 1909. In his classification, **xerophytes** are plants which show morphological and physiological features which could enable them to survive in extremely arid areas. This is accomplished by one of two main strategies, either drought evasion or drought endurance. Drought evaders are the **ephemerals**, the desert plants which complete their life cycles very quickly after rain has fallen and then survive as seeds until after the next storm. Drought endurers can be of many types—for example, storing water as succulents like the cacti, or having deep roots to reach underground water like the acacias. Some other species may rely on physiological endurance to desiccation, such as the creosote bush in the deserts of the United States. This plant can suffer great water loss without damage. Typically, xerophytes have features which could possibly reduce transpiration, such as thick covering layers, hairs on their surfaces, sunken stomata or rolled leaves. The last three characteristics help to establish a micro-climate of high relative humidity round the stomata and prevent water loss by transpiration.

Halophytes, which are plants tolerating saline conditions, have many features in common with xerophytes. They are often succulent and frequently have profuse hairs. In their case the drought is physiological; the presence of salt makes the absorption of water very difficult.

A few flowering plants are adapted to living submerged in water; these are the **hydrophytes**. They have no need for mechanical support but have large air spaces in their tissues to keep them buoyant.

Besides these rather extreme types, Warming recognised a broad unspecialised group of plants called **mesophytes**. These are not adapted to particular water regime environments but still display a wide range of tolerance.

Distributions of plants may largely depend on the effectiveness of precipitation; this will be a function of the kind of precipitation, the type of vegetation present and the rate of evaporation. In many areas fog or dew is important in providing essential moisture for plant growth and thus extending the distribution ranges of species. Drizzle or gentle rain is usually more effective than a hard downpour which will run off the surface quickly. Vegetation will intercept precipitation and decrease the amount reaching the soil where it can be absorbed by the roots. The amount of precipitation lost by **interception** varies with different sorts of vegetation (Fig. 20.4).

Many people have tried to work out the balance between evaporation and precipitation but by far the most successful attempt was made by Thornthwaite in 1948. He put forward the concept of **evapotranspiration**. Potential evapotranspiration (P/E) is the amount of moisture which would be evaporated from the soil plus the water transpired from vegetation in a unit area if the water was sufficient for this to take place to its maximum extent. Drought can be thought of as the condition arising when the amount of water required for P/E exceeds that available from the soil (Fig. 20.5). This approach is very useful in determining where vegetation is limited by water supply.

In the case of animals, water usually only acts as a limiting factor when it is in short supply. There is a great variation in the amounts of water needed for different species but usually cold-blooded animals require less than warm-blooded ones, which use it for heat regulation. Some animals display specific adaptations for survival in arid habitats. Desert animals may avoid the hottest and driest season by becoming inactive—that is, **aestivating**. Others may not avoid the adverse season entirely but adapt by alternative behaviour patterns, perhaps being nocturnal or by burrowing. Ground squirrels in deserts establish micro-climates with high relative humidities in their burrows and so avoid water loss.

Many insects conserve water by secreting semi-solid uric acid instead of urine. This strategy is also employed by desert rodents such as the kangaroo rat, which produces highly concentrated urine and so saves water. The camel

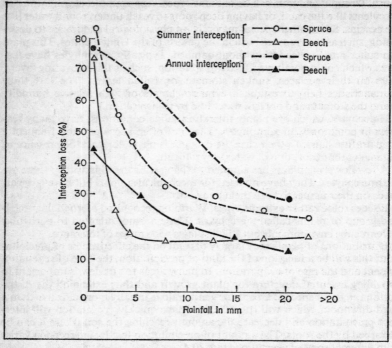

Fig. 20.4. Interception losses from beech and spruce forests.
(Courtesy McGraw-Hill)

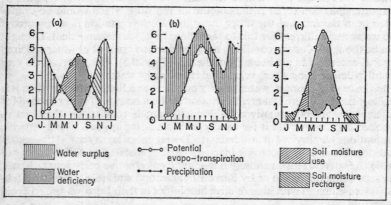

Fig. 20.5. Annual precipitation and potential evapotranspiration at selected stations: (a) Seattle, Washington; (b) Brevard, North Carolina; (c) Grand Junction, Colorado.
(After J. Tivy, courtesy Longman)

is probably the most famous desert endurer yet zoologists still do not really know how it survives. Current opinion is that the main mechanism is simply a physiological endurance to desiccation rather than a water storage or fat storage.

Wind

Wind can act as an environmental factor either directly by causing mechanical damage to plants or indirectly by affecting relative humidity and evaporation rates. High wind velocities can cause an appreciable increase in the rate of transpiration and limit plant growth. In very exposed situations such as mountain summits, coasts and open plains vegetation may be dwarfed as a result of wind action.

Topography

Topography can influence ecosystem development in three major ways. First, by the direct effects of altitude on temperature. Temperature decreases as altitude increases either at the dry adiabatic lapse rate (10°C/km) or, more usually, at a lower rate than this, approximately 6·5°C/km (Chapter Eleven). Second, topography can act indirectly, since temperature changes affect relative humidity. The combination of changes in temperature and relative humidity leads to the development of an altitudinal zonation of ecosystems. Fig. 20.6 shows an example of this from San Francisco Peak, Arizona. At a low level, desert merges into pine forests, which are succeeded by fir and spruce, and then by alpine communities at the highest altitude.

The third way in which topography can influence ecosystem development is by local variation in slope orientation and angle. South-facing sides of valleys receive strong incident light (in the northern hemisphere) and are therefore warmer and drier than north-facing slopes which are in the shadow for a lot of the time. This leads to great contrasts in species structure and productivity between sides of valleys. Angle of slope will be a critical factor in soil formation and drainage.

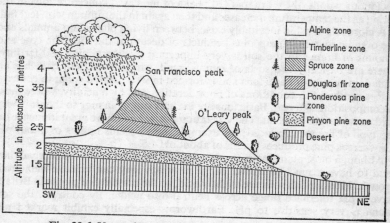

Fig. 20.6. Vegetation zones on San Francisco Peak, Arizona.
(From D. Watts, *Principles of Biogeography*, copyright McGraw-Hill)

Edaphic Factors

The soil is a vital component of terrestrial ecosystems, particularly in cycling nutrients without which all life would cease. Soil and the rest of the ecosystem are closely related; one will influence the workings of the other. Soil-forming processes and different types of soil have already been discussed in Chapter Eighteen; here we need to consider how soil will affect ecosystem development. Particular attributes of soils, such as texture, pH, soil climate and organic content operate in a closely interrelated fashion to exert control on rates of decomposition, nutrient cycling and plant distribution and productivity.

Soil texture is very important in determing the soil climate, since it affects aeration, drainage and ease of root penetration. If the soil is not waterlogged, the pore spaces will contain air which normally has a high concentration of carbon dioxide. This is a result of the respiration of the soil animals without compensating photosynthesis. The concentration of carbon dioxide is an important factor because it influences the pH of the soil, it is a source of carbon for autotrophic microbes, and it may inhibit heterotrophic organisms such as earthworms. High concentrations of carbon dioxide may be more limiting than low concentrations of oxygen. The actual level of concentration will depend on many things but particularly on the soil fauna, the aeration of the soil and the temperature. There will be a gradual outward diffusion of carbon dioxide from the soil and an inward diffusion of oxygen. The rate of this movement will depend on the texture of the soil and the amount of leaf litter on the surface. The soil fauna will be mainly composed of cold-blooded animals whose rates of activity will vary with temperature; as temperature increases their respiratory rate will also increase, producing more carbon dioxide. Similarly, if food (organic material) is abundant, the numbers of soil fauna will increase, resulting in a parallel increase in carbon dioxide production. Consequently there will be seasonal patterns of carbon dioxide concentration reflecting temperature and the availability of organic material. In deciduous woods, there are marked peaks of increased concentrations in the spring as the temperature increases and then again in the autumn with leaf fall.

A close relationship inevitably exists between living plants and animals and the organic matter (humus plus products of decay) in the soil. The type and amount of humus in the soil is very important in determining soil fertility. There are basically two sorts of humus: **mull**, which is nutrient rich, has an abundant supply of basic ions and is formed in well aerated soils; in contrast **mor** is produced in conditions of poor aeration and high acidity, where most decomposition is fungal. Basic ions are in short supply in mor so that it is very acid and nutrient deficient. The vegetation present will have great influence on the type of humus produced; for example, beech yields humus of about pH 6·6 whereas pine produces humus of about pH 4·5.

Although most plants have a wide range of tolerance to pH for survival they tend to have a much narrower range in which they can achieve maximum growth. Most British soils are slightly acid, so the majority of plants in Britain have a pH tolerance range across neutral into acidic. By contrast, the soil fauna is very sensitive to pH. Earthworms especially exhibit marked pH preferences.

One very important influence on soil acidity is the calcium content. Calcium is a dominant ion in soil and gives a physical and chemical stability to the clay humus complex. The presence of calcium ions in the soil solution antagonises the uptake of other ions such as boron, magnesium and phosphate. Because of this, some plants cannot live on calcareous soil and patterns of species distribution arise in relation to the calcium carbonate content of soil. Plants which cannot tolerate calcareous soil are called **calcifuges** and those which are restricted to calcareous soil are called **calcicoles**. Evidence for the distribution of these two types has been collected in the British Isles since the eighteenth century, but experiments to find out exactly why these patterns occur have only been conducted since the 1930s. It is now realised that plants may be restricted to calcareous soil either because of its chemical nature (chemical calcicoles) or because of its physical nature (physical calcicoles). Many traditional calcicoles such as the kidney vetch can, in fact, grow on acidic soil but are usually excluded because of the competition from other plants. A lot of species which act as calcicoles in the British Isles are found on different soils throughout the rest of their ranges. In Britain these species are at the limit of their distribution so that environmental factors become critical and limit them to calcareous soils.

Biotic Factors

Biotic factors are the interactions that occur between living things. Some species are beneficial or even essential for the existence of others, whereas some may be detrimental. Biotic factors are usually far more diverse and intricate than other environmental controls because they rely on the activities of a wide variety of organisms.

Most habitats can be occupied by many different types of plants and animals. The success of a particular species will depend on its ability to obtain its requirements for life. Competition arises if the resources of a habitat are insufficient to meet the demands of all the organisms living there. Generally competition is most intense between individuals of the same species or of different species that have similar ecological niches, especially at young stages in the life cycle. Species will have their maximum competitive ability at their optimum locations but even so, they may be relegated to a sub-optimum location by competition from more vigorous species.

Competition Between Plants

One of the main forms of competition between plants is for light. The dominant plants will be those which grow tallest and modify the light conditions for the rest of the community. They will shade plants growing beneath them and exclude species which require high light intensities. The struggle for light above ground will influence root development and the competition for water and nutrients in the soil. Root competition varies with the vertical and horizontal extension of root systems and the speed at which they develop. Competition will be most severe when plants are drawing from the same level in the soil, and minimised if plant roots are layered. Inability to compete for nutrients and water increases the liability of being shaded out by more aggressive plants.

The mode of reproduction of plants is an important aspect of the competitive ability of species. **Asexual** reproduction (by bulbs, rhizomes, etc.) is a safe way of maintaining a place in the habitat. It can lead to an aggressive spread of a species through an area by excluding other plants from the same community layer. **Sexual** reproduction by seeds is far more precarious because seedlings are very vulnerable. Each seed contains a food store to nourish the young plant until it can make its own food. At one extreme, seeds can be large, containing a lot of food, but these tend to be heavy so they cannot be dispersed far and few can be produced. At the other extreme seeds can be very small, having little stored food, but they can be widely dispersed as they are light and production can be prolific. The small seeds individually have less chance of survival, but because there are many of them it is likely that some will find suitable habitats. The most successful species will be ones which have vigorous, tolerant seedlings which are widely dispersed.

Interactions Between Plants and Animals

Besides competition between plants, biotic factors include the activities of animals which are dependent on plants. The effects of animals are primarily direct. Many plants rely on animals for pollination and seed dispersal. Conversely, many animals are directly dependent on plants for food. Under natural conditions there will be a balance between the biomass of autotrophs and herbivores, so that particular plant species are rarely excluded from vegetation communities. Animals frequently demonstrate scales of preference for different plants and this may lead, for example, to the development of species zonation round rabbit burrows due to selective grazing. By far the most outstanding effects of grazing have occurred where it has been intensified by human activity. This has been responsible for the maintenance of large areas of grassland that would otherwise be forested, such as the chalk grasslands of England.

The resistance of a plant to grazing pressure is related to its life form. Tall plants are quickly grazed out whereas plants which have their buds close to the ground or which grow from the base have a relative advantage.

Interactions Between Animals

The number of animals in any population oscillates between an upper limit which is determined by the energy flow through the ecosystem and a lower limit which will be the minimum number necessary to continue the species. It is ecologically advantageous if the oscillation is controlled in such a way that the danger of extinction is minimised.

There are two basic sorts of controls, either **density dependent** or **density independent**. The latter group includes all the environmental factors which are not influenced by the density of the population, particularly climatic controls. Density-dependent controls operate in relation to the numbers present. Of these, predation is probably the most frequent control of animal numbers. It is a complex process as predators usually prey on several species, and each prey species may have several predators. If the predator's preferred food becomes scarce it will switch to an alternative, and the preferred food species will have a chance to recover its numbers. In this way predation rarely causes extinctions. Many animals have evolved ways of defending themselves from predation such as having camouflage, armour or stings. These defence mechanisms may fail if predators are prolific compared to prey.

A frequent density-dependent control is shown by territorial animals. These require definite territories for hunting and breeding. Weak individuals will be unable to establish territories if there is competition for space and so will be excluded from reproduction.

Man as a Biotic Factor

Man is by far the most important biotic factor. He has caused fundamental modifications of ecosystems by fire, hunting and agriculture. More recently, since industrialisation and the intensification of agriculture, man has obliterated large areas of natural systems and caused pollution of both terrestrial and aquatic habitats. Further consideration is given to this theme in Part Four of this book.

Suggested Further Reading

Ashby, M., *Introduction to Plant Ecology*, Macmillan, London, 1961.

Colinvaux, P., *Introduction to Ecology*, John Wiley & Sons, New York, 1973.

Tivy, J., *Biogeography: A Study of Plants in the Ecosphere*, Oliver & Boyd, Edinburgh, 1971.

Watts, P., *Principles of Biogeography*, McGraw-Hill, London, 1971.

CHAPTER TWENTY-ONE

DISPERSAL AND ADAPTATION

One of the most striking aspects about plants and animals is the diversity of organisms in existence. In both kingdoms there is a vast range of types from the very simple to the highly complex, incorporating a tremendous variety of ecological roles, environmental tolerances and life-forms. Except in the most extreme habitats, life is found throughout the land and aquatic areas of the world and shows great contrasts in the way species are distributed.

Naturalists have long posed the questions, where did all the types of plants and animals come from, how long have they existed, and why are they distributed as they are? Until the nineteenth century the answer to these questions tended to be that all organisms had a divine creation and that species did not change, but in the 1850s two eminent naturalists, Charles Darwin and Arthur Wallace, made a revolutionary contribution to the study of life. They argued that all plants and animals had arisen from pre-existing forms and that they had developed from them by a process of organic evolution. Darwin's and Wallace's ideas have been supplemented and modified by subsequent work, particularly in the field of genetics, so that now we are able to explain the origin and distribution of species more fully.

Sources of Variation

The term species is normally taken to mean any population capable of interbreeding to produce fertile offspring. If we are to understand how one species could evolve from another it is essential for us to look at the characteristics of populations and the variation that occurs within them.

In any population of a species there will be variation in the appearance and physiology of individuals. No member will be exactly the same as another. The potential characteristics of each organism are fixed by its genetic material. This is inherited from the plant's or animal's parents and consists of long protein strands called **chromosomes** which are present in the nucleus of every cell. Each chromosome has **genes** along its length which are formed from molecules of **deoxyribonucleic acid (DNA)**. The DNA is the blueprint for the development of the individual. Every characteristic of a plant or animal's body such as hair colour, leaf shape and flower colour is determined by the DNA. Each gene has a specific role, although several may function together to influence a single characteristic or activity. The chromosome number is usually specific for each species but varies considerably throughout the plant and animal kingdoms. Generally the simple organisms have fewer chromosomes than the species with complex bodies. In higher plants and animals the chromosome number of the main body is called the **diploid number**; at sexual reproduction when the sex cells are formed this diploid number is halved **(haploid)**, so that when two sex cells fuse to form a new individual the correct chromosome number (diploid) is restored. This means that the genetic material of the offspring will be partly derived from one parent and partly from another, resulting in a new combination of genes.

Further variation in combinations or the nature of genetic material can occur through **mutations**. These are induced by radiation or sudden changes in environmental conditions, particularly temperature, and can involve individual genes or whole chromosomes. If the genes mutate there will only be a slight change in the genetic material but if chromosomes mutate, changes will be more drastic. Parts of one chromosome may break off and rejoin with another so that at reproductive stages genes are in different assortments. If the chromosome number fails to be halved at sexual reproduction the number of chromosomes in the next generation will not be the usual diploid but could be a multiple of this becoming **polyploid**. This will introduce more variety to the population.

The genetic material possessed by an individual is known as its **genotype**. This fixes the potential development of an individual but the actual physical expression of this, the **phenotype**, may be modified by the environment. For example, a plant may carry genes to enable it to grow tall but if environmental conditions are adverse it may only achieve short growth. Therefore, the apparent variation in a population is partly due to genetic material and partly due to environmental factors.

Natural Selection

Darwin and Wallace realised in their work of 1858 that variation was the key to evolution. Working independently on island floras and faunas, they arrived simultaneously at the same conclusion, that environmental forces acted upon population variation to bring about an overall change in the characteristics of a species. Darwin and Wallace were both impressed by the fundamental fact that all organisms produce vast numbers of offspring but the populations of species do not keep on increasing. This means that there is great mortality of the young as relatively few survive through to maturity. We have seen that individuals vary slightly in their characteristics, and therefore the organisms which do survive will tend to be the ones which are best suited to the prevailing environmental conditions. The natural habitat is thus exerting a selective pressure on the population. This process of **natural selection** will lead to the persistence of favourable characteristics possessed by the species. The surviving organisms will have genetic combinations suited to the environment, both in terms of tolerance and competitive ability. Their genetic material will be inherited by their offspring, whereas unsuccessful individuals will be eliminated and their unfavourable genetic combinations will be lost from the population. We have also noted that genetic material is not static but can mutate to bring new variation to the population. Natural selection is thus not limited to a sorting of the genetic material at any one time but is a continuing process as new variety emerges.

The total genetic material contained in a population is called a **gene pool**. For each characteristic of the population which exhibits continuous variation, such as height or leaf size, the number of individuals at each height or with each size of leaf tend to follow a particular pattern, known as a **normal distribution curve** (Fig. 21.1). For example, in any population most individuals would be of moderate height for the species, close to the mean for the total population, but there will always be a few members who are very short or very tall. If environmental conditions are stable (Fig. 21.1a), then natural selection will tend to eradicate the extremities of the normal distribution curve, selecting for survival those which possess the average of the particular characteristic and the gene pool will be stable. However, if the environmental conditions are changing (Fig. 21.1b), individuals at one extremity of the normal distribution curve may be more suited to the new conditions and more of these individuals will survive to reproduce. At the same time organisms at the other tail of the distribution curve would be less favoured by environmental change so they would tend to decrease in numbers. This selection due to environmental change results in a shift of the population average as the species adapt and evolutionary change takes place.

Natural selection also occurs in characteristics which do not vary continuously but exist as alternative forms; for example, different colour flowers, or light or dark bodies. In these cases selection operates for definite alternatives and can be seen as a change in the relative numbers of the different types. Many examples of this are observed and it would be useful to look at one instance in detail.

An Example of Natural Selection

Particularly clear examples of natural selection occur when environmental conditions change rapidly, as when pollution affects the habitat of organisms. A very well documented case is that of the peppered moth (*Biston betularia*), which was well known to amateur collectors in the nineteenth century. Until 1845 all known specimens were light in colour, but in that year one black moth of this species was collected from the industrial area of Manchester. The

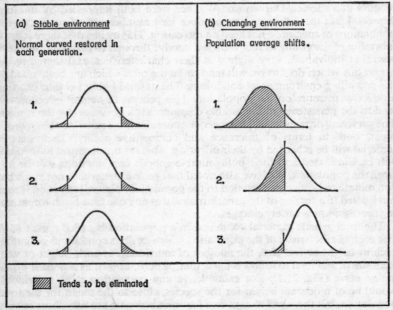

Fig. 21.1. The operation of natural selection on characteristics displaying continuous variation.

frequency of the dark form in the population could not have been more than one per cent at this time. In subsequent years more black peppered moths were found, until by 1895 the dark form comprised at least 99 per cent of the Manchester population. This change corresponds closely to the spread of industrialisation and the parallel spread of air pollution which darkened tree trunks. In pre-industrial times the light forms were well camouflaged against the pale bark of trees and therefore had a distinct advantage over dark forms, which would be easily seen by birds and be eaten by them. Any mutation which produced a dark moth would soon be selected out of the population. As air pollution became severe the dark forms were in an advantageous situation because they were the ones which were camouflaged against the smoke-darkened tree trunks. This phenomenon, known as **industrial melanism**, is also found in many European industrial areas.

Adaptive Radiation

The operation of natural selection may lead to differentiation of sections of the population over the area occupied by the species. Local environmental conditions may select characteristics which would not be favoured elsewhere so that the individuals become adapted to their particular surroundings. If a species is spreading out over an area it will probably encounter a variety of conditions which necessitate a corresponding variety of adaptations. Changes induced by spread in this way are said to be caused by **adaptive radiation**.

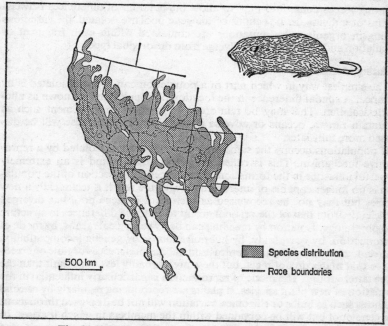

Species distribution

Race boundaries

500 km

Fig. 21.2. Ecological races in the western pocket gopher.

Many examples of this can be seen both for species and for larger taxonomic groups.

At the species level adaptive radiation can lead to the formation of **races** or **ecotypes**. Ecotypes within the species may exhibit different tolerance ranges or slight differences in appearance but they will still be capable of breeding with the rest of the individuals of the species population. An interesting example of this is shown by the western pocket gopher (*Thomomys bottae*), which inhabits the south-western area of the United States and is capable of adaptation to a wide variety of soil conditions, foods, temperatures and relative humidity. Consequently a large number of races within the species have developed; these can be distinguished on the basis of size and coat colour. Each race occupies only a small section of the total species area.

The Formation of New Species

So far we have considered how the characteristics of a population could change by natural selection within a species, but we have not yet answered the question of how a completely new species could arise. Obviously, if inter-breeding occurs between members of a population the genetic material of the group will be mixed; any tendency for variation will be dispersed through the gene pool. In order to designate a population as a new species it has to be in-capable of breeding with the type from which it has originated. If change occurs slowly in a population through time it may adapt to different environ-mental circumstances and evolve to a form which contrasts with its original. In this case, populations would be separated by time, but more usually other isolating forces come into play to allow the build-up of differences between groups of individuals. If sections of the gene pool are isolated, the variations arising in organisms by mutations are contained within each fragment of population and their genotypes diverge from the original type.

Isolating Mechanisms

The simplest way in which part of a population can become isolated is by distance. A spatial difference in the location of populations is known as **allo-patric isolation**. This may be reinforced by barriers to movement such as mountain ranges, oceans or adverse climatic conditions. These will be dis-cussed more fully later.

If populations occupy the same area they can become isolated by a repro-ductive mechanism. This is called **sympatric isolation** and is an extremely powerful influence in the formation of new species. If a section of the popula-tion is no longer capable of interbreeding with the rest it is technically a new species but may not be recognised as such until its gene pool has diverged sufficiently from that of the original group to produce differences in structure and physiology. Isolation by breeding can be by ecological means, by mode of reproduction, by polyploidy, by hybridisation or by genetic incompatibility. The ecological means include contrasts in habitat preferences or time of breed-ing so that mates do not meet, but these are probably less important than the three latter groups. The mode of reproduction is particularly influential in the formation of new plant species. If plants are reproducing regularly by asexual methods such as bulbs or rhizomes, variation will not be dispersed throughout the gene pool but will be contained within the members in which it arises. In this case it is easy to envisage divergence within the species in response to the variation of environmental conditions that may be encountered over the area occupied. However, if plants usually reproduce sexually, isolation is more difficult. It may arise by mutations such as polyploidy. Polyploid individuals are usually larger and more vigorous than diploids, and so frequently they have a competitive advantage. Many present-day plant species, especially agricultural crops such as wheat and cotton, are in fact polyploids derived from pre-existing diploid forms. Polyploids cannot mate back with diploid types. They may retain fertility for sexual reproduction but sometimes they have to resort to asexual methods if they are to persist.

Hybridisation, which is the result of mating between two different species, is also a frequent originator of new species. Hybrids are common as reproductive isolation of a species is rarely 100 per cent effective. Closely related species are especially prone to crossing but their offspring are usually infertile. This is because they have two completely different sets of chromosomes (one set from each species) and it is difficult to divide the genetic material in the way required for sexual reproduction. Hybrids can survive as a new species only if they can reproduce asexually or if their fertility is restored. This can be done if a mutation occurs to produce a polyploid from the hybrid, so that the chromosome number is multiplied and there are at least two complete sets of genetic material which allow the division required for sexual reproduction to take place.

Genetic isolation may result from mutations which render sex cells incompatible so no fertilisation occurs, or if fertilisation does take place the resulting embryo does not develop properly. This effectively prevents interbreeding. Change in the spatial location of populations has also been an extremely important factor in species formation and this leads us to consider migration and dispersal in some detail.

Migration and Dispersal

Migration can be defined as the spread of a species into a new area and usually implies movement over long distances through time. Both plants and animals can migrate, although in the case of animals it may be of a temporary nature for breeding or to avoid adverse climatic conditions at one season of the year. The ability to migrate, whether on a permanent or temporary basis, is an important aspect for the survival of a species because it allows adjustment of the locations inhabited if climates change (Chapter Seventeen), and also facilitates the extension of the area occupied if population pressure or severe competition from other species builds up.

Dispersal in Plants

Migration requires a change in the location of the individuals that make up the species. For plants this will obviously only be possible at the reproductive stages, the position of the offspring being different from that of the parents. Therefore the speed of migration and hence the ability of a plant species to keep pace with climatic change will be closely linked to life cycles and methods of reproduction.

Plants which reproduce vegetatively by asexual means, such as rhizomes, corms and bulbs, will have an extremely limited capacity for migration as the offspring will be immediately adjacent to the parents and so will hardly be dispersed at all from the previous generation. In contrast, plants which reproduce by spores or seeds will have faster rates of migration as they will be dispersed further between generations. The seed plants are by far the most efficient at dispersal and exhibit a vast range of adaptations to aid movement away from the parent plant. Most of these species rely on external agents, although there are a few which have exploding or twisting seed cases that expel the seeds. In other species the seeds may not have any structural modification but may be light enough to be blown about by the wind; a good example of this can be seen in the dust seeds of the orchids, which are minute but are produced in vast numbers. Many seeds have special adaptations such as wings or plumes of hairs to help them to become dispersed by the wind, whereas others may rely on water or even ice for transport. There are also many examples of seed dispersal by animals. Hooked or sticky seeds may adher externally to fur, and others may be carried in mud clinging to feet. In addition, seeds may be dispersed by being eaten, particularly by birds, and then passed out with the faeces of the animal. Humans have contributed to the dispersal of plants in similar ways to these. Many agricultural pests are the result of the unwitting transport of seeds by humans travelling from one area to another.

We have already noted that the possibility of dispersal will only arise for plants at reproduction; therefore the longevity of the species and the time taken to reach maturity are also important aspects to consider. **Annual** plants have a distinct advantage here because they mature to reproduce within one year and consequently possess the potential for migration every year. Plants that require two years to complete their life cycle, the **biennials**, will only be able to migrate every two years, and the **perennials**, which may live for scores of years and may not mature until the plant is fifteen or so years old, will have far less frequent chances of migration.

Dispersal in Animals

Animals generally have far more capacity for dispersal than plants because most of them are mobile, and they can also be aided in their movement by external agents. The agent can be biotic or abiotic but in either case the resulting change in location is said to be due to **passive dispersal**. Parasites and symbionts are carried with their hosts, whereas many other animals are blown about by the wind—particularly the insects—or float with water currents.

If a change in the location of the individual happens because that animal has moved itself, it is called **active dispersal**. The rates of dispersal and hence migration in this case will depend on how fast the animal can move about. At one extreme there will be very slow moving diggers and crawlers, such as the earthworms, and at the other extreme, there will be fast runners, hoppers and flyers which can move long distances in short times.

Migration in Plants and Animals

There are three main patterns of migration in plants and animals. First, there can be an extension of the area occupied (the geographical range of the species); second, expansion can take place in one direction only while contraction is taking place in other directions so that there is a general shift in the area of the location inhabited; and third, a situation could arise when the range of the species is contracting on all sides at once. This last case is not technically migration but when we consider evolution and the pattern of present species distributions, it is useful to think of this condition as **retreating migration**.

It is the first case which gives rise to adaptive radiation; if a species is migrating without a shift in the climatic zones it will be necessary for organisms to adjust to new conditions encountered. The second situation usually arises in response to climatic change, when species migrate parallel to the environmental conditions they tolerate. Although the species themselves may not change, the structure of the vegetation and animal communities may alter drastically. We have seen that species migrate at different rates so the communities do not migrate as a unit. This leads to great reassortment of the species living together and consequently the floras and faunas of a region are made up from many components which show affinity to different areas and thus provide evidence for past migrations. Retreating migration occurs if conditions become unfavourable for a species on all its boundaries, possibly as a result of inability to keep pace with climatic change. Zones of sterility develop round the periphery of the geographical range, and gradually the area occupied contracts. A few individuals may survive as **relicts** in locally favourably areas known as **refugia**. Some examples of relicts and refugia will be given in the next chapter.

Barriers to Migration

Although plants and animals may possess the capacity for migration it may not always be possible for them to do so. They may encounter barriers. These could take the form of competition from the fauna and flora already occupying the area, but the success of the migrating species in comparison with the resident species will depend on their relative tolerance ranges and demands from the environment. Alternatively, the barriers to migration may be due to abiotic environmental factors. This will be especially apparent if the climates are not changing. A species may migrate to the limits of its tolerance range and then no further without adaptation. In addition there may be actual physical barriers that impede dispersal—for example, oceans or lakes. The effect of these as barriers will of course vary with their size. Time can be considered as a barrier to migration as well, because slow-spreading species may be overcome by changing conditions, and also the length of time a species faces a barrier is very important. This is because natural selection may take place in response to the stress imposed by the barrier, leading to evolutionary changes which might enable the species to continue to migrate.

The Pattern of Evolution

Earlier in this chapter we saw how new species can develop from pre-existing ones. We must now consider the events that have taken place in the past to produce the diversity of life as we know it, so our attention will be directed to the evolution of the main groups of organisms. The basic features of speciation are fairly well understood but the complex processes leading to the appearance of different forms of life are known only in the crudest outline. The interaction of variation and natural selection and the development of new adaptations must be very important in these processes, but change at the species level does not seem to be sufficient to explain the emergence of new biological systems. The history of the major changes in life suggests that other factors are operating above the species level. It would be useful at this point to examine the trends in evolution in more detail.

Trends in Evolution

Every geological period has been dominated by one particular form of life; as conditions changed to those of a new period, different types of organisms emerged to dominance. Within each period, natural selection has created a vast array of adaptive experiments culminating in a breakthrough to a new biological system that may cross the ecological barrier to the next period. Every breakthrough or shift in dominance is dependent on ones that have occurred before so that the process of evolution is progressive. Individual shifts rely on opportunity, ecological access, and the development of a new adaptation. Characteristically the shifts are rapid and are followed by a bout of adaptive radiation into the new ecological zone—for example, with the emergence of life on land. The origin of completely new biological systems is comparatively rare, so consequently only a few major types have developed during the history of life. It is interesting that although many species have become extinct none of the major groups or **phyla** (singular: phylum) has disappeared.

One very marked evolutionary trend illustrated by many examples in the fossil record is the progressive sustained tendency for characteristics to develop along certain lines. These long-term continuing trends rarely appear in only one structure but usually involve a complex of features. A classic and well documented example of this can be seen in the evolution of the horse (Fig. 21.3). The evolution of the modern horse (*Equus*) from its small Eocene an-

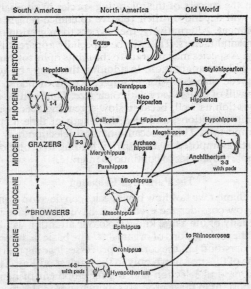

Fig. 21.3. Evolution in horses from the Eocene to modern forms. Numbers refer to toes: ancestral type has four toes on front feet and three on back feet, padded as in dogs. Modern form has only one toe on each foot.

cestor (*Hyracotherium*) was not direct along a single evolutionary line but involved many intermediates and associated forms. Although Fig. 21.3 is greatly simplified it does show that development was by continuing trends along certain lines, particularly in the reduction of the number of toes, the increase in body size, and the relative lengthening of the legs. All of these trends developed in response to the change of environment from that of lush vegetation which could be browsed, to a drier grass habitat that necessitated grazing and speed of flight from predators who could easily see the horses in the open.

The Evolution of the Main Groups of Plants and Animals

It is now known that life has existed on earth for at least 2500 million years. The earliest forms of life recognised are algae and very simple animals of the Precambrian. The complexity and variety of life increased slowly at first from these beginnings, accelerating as environmental changes took place and diversity developed.

Fig. 21.4 illustrates the major developments through time of groups of animals and plants that have fairly good fossil records. Reference to this shows that by the end of the Cambrian most of the major groups of **invertebrates** had evolved. There were sponges, coelenterates such as jelly fish, corals and sea anemones, and early types of snails. By far the most abundant type in the Cambrian, however, was the trilobite. This looked like an aquatic wood louse, and persisted on into the Permian but became less important as other types emerged. Simple sorts of plants such as the **Thallophytes** (the algae and fungi), the **Bryophytes** (the moss and liverwort group), and primitive plants called **Psilophytes** were present in the Cambrian but were not very diverse.

As the trilobites declined, armoured **Eurypterids**, which are related to modern scorpions and could be as large as six feet long, came into dominance. These remained in evidence to the Silurian and Ordovician, together with a range of other invertebrates such as snails, worms and creatures similar to the modern octopus. At about this time two important developments show themselves in the fossil record. First, there are the first land plants, developing to greater prominence in the Devonian. In this period there are prolific records of club mosses, ferns and seed ferns. Second, the first **vertebrates** developed. These were jawless fishes, some of which were heavily armoured. Shortly after the emergence of this new biological type, scorpions, spiders and insects also developed, although the insects had not yet evolved the power of flight. Fish became more abundant as adaptive radiation took place, and became the most prominent type of the period. Because of this, the Devonian is known as the age of fishes.

By the Carboniferous new groups of vertebrates had evolved and invaded the land habitats; these new groups were the **amphibians** and **reptiles**. The reptiles were particularly successful and dominated the land for a long time. Plant life had developed to form large coal-producing forests and swamps by this period, and provided an ideal environment for the reptiles. Many forms developed within the reptile group; there were great contrasts in size, locomotion, food preferences and habitats. The age of reptiles reached its climax in the Jurassic, then the group suddenly lost its superiority in the Cretaceous. This was most probably due to a change in climate as many other types also declined with them, particularly the primitive seed ferns and the trees of the Carboniferous forests.

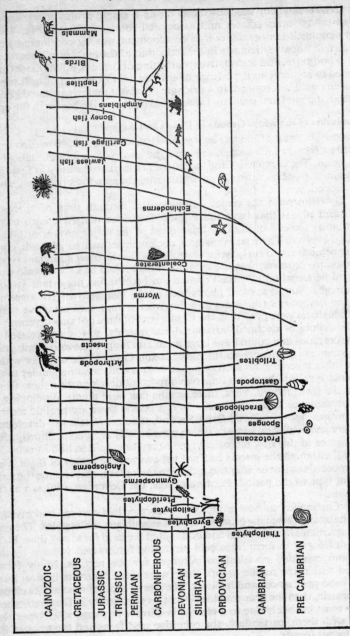

Fig. 21.4. Geological time chart showing the emergence of the major types of plants and animals.

The ecological void created by the decline of these groups presented other forms with the opportunity for development without severe competition. Flowering plants became prolific and by the middle Cretaceous were not very different from those existing today. **Birds** and **mammals** had been evolving during the age of reptiles and towards the end of the Cretaceous they diverged rapidly in an evolutionary spurt fostered by the favourable climatic and vegetational conditions. Both of these animal types were highly successful and have dominated the land ever since.

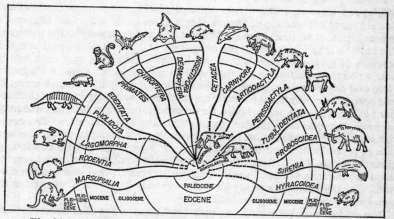

Fig. 21.5. The main evolutionary lines in mammals since the Eocene.
(From W. George, *Animal Geography*, Heinemann)

By the beginning of the Tertiary all the main types of mammals had evolved. Subsequently there has been divergence within these main orders to produce a tremendous variety of species, and the distribution of the orders has been affected by continental drift and climatic change, as outlined in the next chapter. Fig. 21.5 illustrates the main evolutionary lines of the mammals in the Tertiary.

Suggested Further Reading

Dowdeswell, W. H., *The Mechanism of Evolution*, Heinemann, London, 1963.
George, W., *Animal Geography*, Heinemann, London, 1962.
Olson, E. C., *The Evolution of Life*, Weidenfeld & Nicolson, London, 1965.
Savage, J. M., *Evolution*, Holt, Reinhart and Winston, London, 1971.
Seddon, B., *Introduction to Biogeography*, Duckworth, London, 1971.

CHAPTER TWENTY-TWO

THE HISTORICAL FACTOR IN THE DISTRIBUTION OF SPECIES

In earlier chapters in this part of the book it was seen that the occurrence of species can be related to a number of biological and environmental factors. Sometimes an explanation of present distributions solely in the light of prevailing ecological conditions does not provide a complete answer to the patterns. We have to take into account the conditions and events of the past and the reaction of species to those situations. Some of the events may have had a bearing on the evolutionary trends described in the previous chapter. Chapters Nine and Seventeen discussed a range of climatic and landscape changes; it was noted, for example, that climate has undergone some very marked fluctuations in the recent geological past. Environmental changes such as these may take some time to exert themselves on plants and animals: species and communities possess considerable inertia, depending, for example, on the average rate of reproduction and the longevity of each generation. With continuing climatic change it is inevitable that some animal and plant species and vegetational boundaries are not in equilibrium with present environmental conditions. Equally, tectonic and sea-level changes may have caused disruption to previously continuous distributions.

The evidence for reconstructing past conditions and species behaviour lies in fossils, particularly the bones and shells of animals, and the pollen, seeds and leaves from plants. Since we are seeking to explain the occurrence of modern species, we are interested mainly in fossils that are the same as, or closely related to, living specimens. In the case of plants, species very similar to today's were in existence by the end of the Mesozoic (Chapter Twenty-One). In the animal world, the same applies to reptiles, fishes and invertebrate animals. But with mammals and birds, the rate of evolution in the Tertiary has been so rapid that close affinity between modern and fossil species in many cases only goes back two million years or so to the early Pleistocene.

The Distant Past

The Effects of Continental Drift

Biogeographers have long been aware that the distribution of certain species, particularly amphibians and reptiles, could only be satisfactorily explained by the existence at some time in the past of land connections between continents. For example, closely related tongueless frogs (*Aglossa*) are found in both West Africa and the Amazon region, despite their separation by thousands of miles of ocean. Up to the 1960s such distributions were often explained in terms of ancient **land bridges**, supposed to have foundered long ago. Nowadays, the framework of plate tectonics (Chapter Two) is seen to offer a much better basis for the facts, and it is accepted that continental drift has been a significant influence on the distribution of major groups of plants and animals.

The effects of continental drift are not the same for both animals and plants.

If we first look at the animal regions of the world (Fig. 22.1), the fauna of South America shows more similarities with Australia than it does with North America, despite the land connection with the latter. This is because the two American continents have only relatively recently been joined, whereas both South America and Australia were once part of Gondwanaland, the southern supercontinent.

The timing of the separation of the major land masses is critical in relation to the rate of evolution of mammals. At the largest scale, this is best seen in the distribution of **marsupials** (whose young complete their development in the mother's pouch). These first appeared in the Lower Cretaceous, before the final break-up of the supercontinents which occurred in the middle part of the

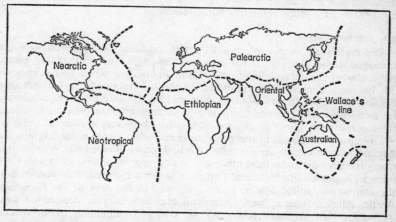

Fig. 22.1. World zoogeographical regions, based on the distribution of mammals.

Cretaceous. Later, with the extinction of the dinosaurs in the Upper Cretaceous, more advanced mammals, the **placentals** (in which the whole period of embryonic development takes place in the uterus) underwent rapid evolution, virtually replacing marsupials in many parts of the world. Had the placentals managed to spread to every corner of the globe before continental disintegration, marsupials would almost certainly have been wiped out. However, Australia became separated from the other land masses *after* it had been colonised by marsupials, but *before* the great evolutionary expansion of the placentals. Since the Cretaceous, Australia has remained unconnected to other land masses and, apart from rats and bats, the continent has no indigenous placental animals. As a result, it retains a marsupial fauna which has developed into a wide variety of forms which occupy the niches that placentals have filled elsewhere. By contrast, in North America both placentals and marsupials existed together at first, but only placentals remained by the Pliocene. In South America, where land connection with North America has been intermittent, a few marsupials still remain.

Since Australia reached its present position, its marsupial fauna has spread westward by island-hopping towards south-east Asia. Conversely, some placental mammals of the Oriental region have spread eastward, and in the

Far East there is now a narrow area of overlap, sometimes known as **Wallace's Line** (see Fig. 22.1).

The floral regions of the world today, based on the distribution of **angiosperms** (flowering plants), show fewer basic differences than the animal regions. Although each each floral region displays individual features at the generic and species level, they are not basically unique in the sense that each has major groups found nowhere else. One of the most fundamental features is that almost everywhere in the world, four families are among the six most numerous: the Compositae, Gramineae (grasses), Cyperaceae (sedges) and Leguminosae. Their distribution suggests that angiosperms spread through the world before the continents split apart, and this is confirmed by fossil evidence which demonstrates that angiosperms originated in the Jurassic and had reached most parts of the globe before the mid-Cretaceous.

The Tertiary Period

The continents continued to drift apart towards their present positions throughout the Tertiary. Other important changes were also taking place, particularly in respect of climate and mountain building, which had an important bearing on present distributions. At the beginning of the Tertiary, the climate of the world was both warmer and much more uniform than now. Many species of plants and animals were circumglobal in their distribution: the Eocene subtropical conditions of London were very similar to those of Washington and China. In more northerly latitudes, forests extended into the Arctic circle, creating the so-called **Arcto-Tertiary forest belt**. Within this, many species which now have relatively limited geographical associations were once widespread. For instance, *Taxodium* (swamp cypress), now confined to the south-east United States, was widespread in the area of the Canadian Arctic, Alaska, Siberia, west Greenland and Spitzbergen. Associated with *Taxodium* were *Sequoia* (redwoods) and *Ginkgo* (maidenhair tree), both of which are recognised as geographical relicts by the peculiarities of their modern distributions. Another example is the Chestnut family (Fig. 22.2). The disjunct distribution of the three species existing today is a relic of much greater ranges in the Arcto-Tertiary forest.

During the Tertiary, there was a gradual cooling of climates in the polar regions, leading to increased differentiation between the faunas of North America and Eurasia on the one hand and the Oriental and Ethiopian (African plus Madagascar) regions on the other. The London Clay (Eocene) flora and fauna shows that tropical conditions still existed in southern England fifty million years ago. Of about a hundred genera of plants preserved in the clay, seventy-three have living relatives, nearly all of which are found in the intertropical zone. Tropical animals such as crocodiles and turtles have also been recovered from London Clay. The Oligocene, Miocene and Pliocene witnessed a gradual increase of forms more tolerant of temperate and cool climates, until by the end of the Pliocene, an essentially modern flora existed in Europe, with 60 per cent of the plants still found within the continent today.

An additional factor which undoubtedly contributed to making later Tertiary floras and faunas considerably less uniform over the globe was the onset of orogenic activity in mid-Tertiary times, the Alpine mountain building period. This was responsible for the uplift of the Alps, the Rockies, the Andes and the Himalayas. These mountains would have broken up the broad forest belts of

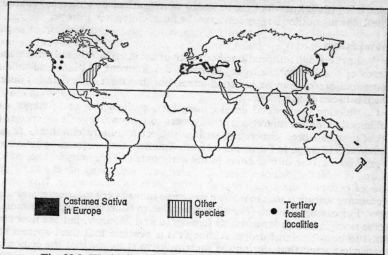

Fig. 22.2. World distribution of the chestnut family (*Castanea*).
(From B. Seddon, *Introduction to Biogeography*, Duckworth)

the early Tertiary into relict pockets and would also have had a strong effect on global climate (see Chapter Thirteen).

Pleistocene Changes

In the Pleistocene, the climate was characterised by alternations of cold and warm episodes, the more recent cycles being marked by the advance and recession of continental ice-sheets (see Table 9.1). Unlike continental drift, these climatic oscillations did not radically alter the major orders of plants and animals already established by mid-Tertiary times, but they did have an important effect on the distribution of species within each of the world's plant and animal regions. In the lists of Pleistocene fossil plants and animals from various parts of the world, there appears to be little evidence of extinction or evolution except among the vertebrates; rather, the geographical ranges of species have often changed markedly.

There are two ways in which Pleistocene climatic change has affected distributions: by directly causing repeated equatorward and poleward migrations of plants and animals with each cold and warm stage respectively; and by indirectly creating and disrupting land bridges through associated glacio-eustatic changes of sea-level (Chapter Eight), as between Britain and continental Europe, and between Asia and North America at the Bering Straits. In the latter case, the existence of a land corridor during glacial episodes and the early part of interglacial periods probably explains why the plant life of Alaska and eastern Siberia is strikingly similar, even to the extent of possessing many identical species.

The technique of pollen analysis has enabled a fairly complete record of Pleistocene vegetational change to be obtained, especially for Britain and north-west Europe. The faunal record is not so well documented: some stratigraphical work has been done on insects, molluscs and foraminifera, but relatively little is known of the sequence of migrations of vertebrate animals. The following outline largely concerns the floral changes in Europe.

Warm Stages

We have enough information about each of the interglacials and other warm stages of the Pleistocene to say that the development of plant and animal communities were similar in broad terms, but that there were detailed differences between each episode.

Considering the general development first, each of the warm stages was sufficiently long for migrational readvances to reinstate repeatedly species in areas that had been either glaciated or subject to tundra conditions in the previous cold stage. In Europe and North America, there appears to have been a gradual increase of tree cover in the early part of each warm stage, culminating in mixed deciduous forest. On each occasion, many plant and animal species occupied the same areas as in the previous warm stage. However, a significant feature is that in many cases these ranges were more extensive than now. For example, beech (*Fagus*) and hornbeam (*Carpinus*) were common in some interglacials as far north as Leningrad and Moscow, but are now confined to non-Russian Europe. Although it is possible that these species had longer to spread than during the present warm stage, or that the ecological tolerance of species may have changed, it is generally agreed that some interglacial periods were warmer than the present one and experienced adequate rainfall for forest to dominate over wide areas. This climatic influence is confirmed by the remains of certain subtropical animals. For example, hippopotamus ranged over the more oceanic parts of Europe in the last interglacial as far north as East Yorkshire.

Although the majority of species have reappeared in the same area in each warm stage, in other cases ecological behaviour has not always been the same, and this provides a degree of contrast between the stages. We can illustrate this with reference to the hazel (*Corylus*). This tree has been present in Britain in at least the last six warm stages, including the present one, the Flandrian. In Fig. 22.3, we may note two points of difference between the stages: first, the

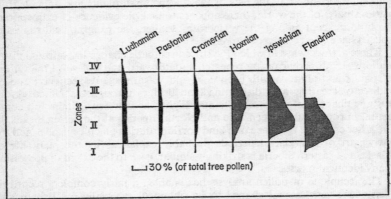

Fig. 22.3. *Corylus* (hazel) pollen curves for Pleistocene temperate stages in southern Britain.

abundance of *Corylus* in the first three stages was much less than in the Hoxnian, Ipswichian and Flandrian; and second, in the latter three, the timing of its expansion became progressively earlier. One possible reason for these divergences may be that the cold stage refuges of *Corylus* have changed through the Pleistocene, allowing the species to reach Britain at different times and with varying abundance in each warm stage. Unfortunately, it is difficult to be precise about the location of the glacial refuges of *Corylus* and other thermophilous genera: they may have been on the oceanic fringes of the continent or in small areas in southern Europe.

Another element of contrast is that there are a number of Pleistocene species which appeared in only one or two of the warm stages. These species contribute some of the most distinctive features of warm stage floras and faunas, and provide a means of identifying the age of different temperate deposits. In ecological terms, these species were important in modifying the appearance and functional aspect of the vegetation, as in the early Pleistocene, when a number of additional evergreen species diversified the deciduous forests. Particular examples of interglacial species once found in Britain but now no longer native to this country include silver fir (*Abies alba*), box (*Buxus sempervirens*), both present in the Hoxnian interglacial, and Balkan maple (*Acer monspessulanum*) which was a component of the deciduous forest of southern England in the Ipswichian (last) interglacial. Silver fir is now a central European montane species, box is most frequently found in southern Europe, and the distribution of the Balkan maple is centred in Yugoslavia, Greece and Albania, as its name suggests (Fig. 22.4).

A few of the species once found in Britain in the early Pleistocene warm stages have now completely disappeared from Europe. In the Ludhamian and Antian stages (see Table 9.1), hemlock (*Tsuga*) and wingnut (*Pterocarya*) were present, but they have not been discovered in deposits of any subsequent stage of the Pleistocene. These two species are now found only in North

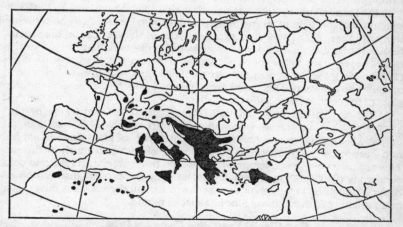

Fig. 22.4. Modern (*black*) and last interglacial (*crosses*) distribution of Balkan Maple (*Acer monspessulanum*).

(From B. Seddon, *Introduction to Biogeography*, Duckworth)

America, where they have survived throughout the Pleistocene. The explanation for their extinction in Europe on the one hand and their survival in America on the other, would seem to lie in the fact that in Europe high mountain ranges run east–west and so may have prevented the dispersal of these and many other species to more southerly latitudes in the cold episodes. Equally, a barrier would have been presented to the northwards return of thermophilous species at the beginning of warm stages. In America by contrast, the high mountain ranges run north–south, allowing free dispersal of species towards the equator during glaciation and during their subsequent return. Consequently, we find that many more species were permanently extinguished from the European flora during the Pleistocene than from North America. Today, there is a much greater variety of forest trees in North America than in Europe at comparable latitudes.

Cold Stages

Until the recent establishment of a detailed Pleistocene biostratigraphy on the basis of pollen analysis (see Table 9.1) there was considerable speculation as to what had happened to the flora and fauna of Britain in the cold stages, particularly during the glaciations. One popular view was that because ice had covered most of the country and the rest was subject to permafrost, nothing could have survived the glacial episodes. The total present-day flora and fauna of the British Isles had therefore immigrated during the post-glacial. An alternative view, which attempted to take into account the present disjunct distribution of Arctic–Alpine plants on British mountain tops (Fig. 22.5), suggested that some species had survived on **nunataks** (areas above the ice) throughout glaciation. However, it is now appreciated that long periods within the cold stages were characterised by tundra rather than glacial conditions. Thus, although trees and other thermophilous plants migrated southward with each cold stage, many herbs, grasses and sedges remained in lowland Britain, and also colonised the exposed floor of the surrounding seas.

In the Devensian stage (the last glaciation) it is likely that plants and animals were present throughout, even during the maximum ice advance about twenty thousand years ago. In the less severe episodes of the Devensian, the vegetation in lowland Britain took on the aspect of a discontinuous mixed sward, treeless except for a few birch shrubs in sheltered localities. Dwarf birch (*Betula nana*), least willow (*Salix herbacea*) and mountain avens (*Dryas octopetala*) were common. This sward provided good grazing for reindeer in some parts of the country, especially the Midlands.

At the close of the last glaciation there occurred a climatic amelioration known as the late-glacial Interstadial (approximately 11,300–9500 B.C.), during which there was a great northward invasion of plant and animal life into Europe. A rich herbaceous vegetation developed over much of Britain. This explosion of life was temporarily checked by a subsequent return to cold conditions (9500–8000 B.C.) but it is clear that the present distribution of Arctic–Alpines and many montane species in Britain derives from their much more widespread extent during the late-glacial period.

Post-Glacial Vegetation Change in Britain

The onset of rising temperatures around 8000 B.C., marking the beginning of the post-glacial period, had a dramatic effect on the vegetation of the country. Periglacial activity ceased, soils became stabilised and a rapid influx of the thermophilous plants commenced across the dry ground of the English Channel and North Sea. The park-tundra conditions of the late-glacial were steadily replaced from south to north by woodland as the climate improved. With the reduction in open ground habitats, many herbaceous species became extinct in lowland areas, surviving in relict pockets only on steep cliffs or above the tree line, where they can still be found today (see Fig. 22.5). Another result of the increase in woodland cover was the disappearance from the fauna of large herbivorous mammals common in the late-glacial, such as the elk, the giant Irish deer and reindeer.

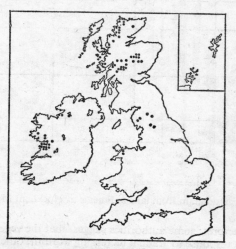

Fig. 22.5. Present-day distribution of a plant widespread in the late glacial: *Dryas octopetala* (mountain avens).

A typical pollen diagram for the post-glacial (Fig. 22.6) shows a well-defined progression in the development of the forest cover in Britain. After a brief increase in juniper, the first trees to make a more or less closed woodland were birches (*Betula*); even now the birch forest goes further north in Europe than other trees. These gradually gave way to a pine (*Pinus*) and hazel (*Corylus*) forest, with hazel either acting as an undershrub or, in some areas, forming pure hazel woods. After about 6000 B.C., familiar broad-leaved deciduous trees such as oak (*Quercus*), elm (*Ulmus*) and lime (*Tilia*) appeared in quantity and began to replace the pine. By the so-called Atlantic period (Table 17.1) a virtually continuous forest covered Britain: in England and Wales this mostly took the form of mixed oak forest, except in the wetter valleys, where alder (*Alnus*) was very abundant. In Scotland, pine forests continued to dominate. This was the period of greatest post-glacial warmth: trees such as *Tilia* achieved their greatest range and abundance, and the altitudinal treeline reached levels considerably higher than now on British mountains.

The precise cause of these forest changes in the early part of the post-glacial

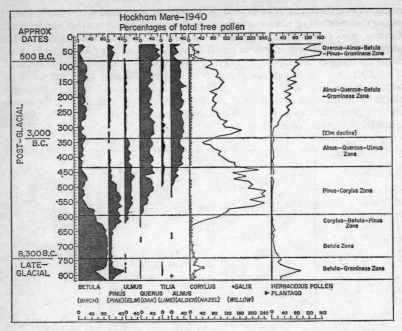

Fig. 22.6. Pollen diagram from lake sediments at Hockham Mere, Norfolk.

is not easy to interpret: some authorities suggest that the vegetational succession represents a response to gradual increasing warmth; others hold the view that optimum temperature conditions for mixed deciduous forest were achieved quite quickly, and that the succession is more related to biotic factors, such as rates of colonisation.

At about 5500 B.C. Britain became isolated from continental Europe by the post-glacial rise in sea-level. For the same reason, Ireland had already been isolated from Britain about a thousand years earlier. These events are significant in explaining certain plant and animal distributions. For example, relatively late immigrants into Britain, such as lime, hornbeam (*Carpinus*) and Herb Paris (*Paris quadrifolia*) are not found in Ireland. Fossil evidence shows that Britain and northern France possessed almost identical floras up to 5500 B.C., but after that, whereas France continued to be enriched by migrating species from elsewhere in Europe, Britain received hardly any, so that now it possesses only about two thirds of the plant species of Europe. In the case of land mammals, the disparity is even more marked: Europe has 167 species, Britain 41 and Ireland only 21. The well-fabled lack of snakes in Ireland can also be attributed to the early isolation of the island from Britain in the postglacial.

One interesting group of plants that did succeed in reaching Ireland in the post-glacial is the Lusitanian element, found in the west of the country. The strawberry tree (*Arbutus unedo*) and giant butterwort (*Pinguicula grandiflora*) are not found wild in Britain: other species such as Cornish heath (*Erica vagans*) and pale butterwort (*Pinguicula lusitanica*) are found in south-west England as well as Ireland. It seems likely that these plants spread from Spain and Portugal up the Atlantic seaboard during the period of low sea-level in early post-glacial times.

We normally regard the species which had reached Britain by the time of severance from the continent as the **native flora and fauna**. Although later natural introductions may have occurred from a small group of species with dispersal mechanisms capable of overcoming the sea-barrier, by far the bulk of later additions are seen as 'artificial', introduced deliberately or accidentally by man, who becomes the dominant factor in the environment in the latter part of the post-glacial.

The Impact of Man

The story of the latter part of the post-glacial in Britain is one of a gradual depletion of the forest cover, largely as a result of man's activities. In addition, a slight deterioration in climatic conditions has probably also contributed to reducing the range of some species. The cooler and wetter conditions which set in around 500 B.C. (Chapter Seventeen) caused widespread blanket bog formation in many upland areas and a drop in the treeline. Buried pine or birch stumps can often be seen at the base of peat cuttings in western Britain.

The first substantial impact of man on the natural vegetation of Britain dates from the arrival of Neolithic man about 3300 B.C. Before this, Palaeolithic and Mesolithic man, living largely by hunting and foraging, had relatively little impact on the environment, but the Neolithic witnessed the introduction of new agricultural practices which had a permanent effect on the natural ecosystems. This interference took several forms. First, wholesale clearance of trees took place as part of a system of shifting cultivation; initially this only occurred on a local scale, but later in the Neolithic, more permanently cleared areas were established. Second, the pasturing of livestock prevented the regeneration of the forest in and near the cleared areas. Third, selective usage of certain tree species was practised; in particular, elm leaves were used as a source of fodder, and an **elm decline** can often be recognised in the pollen diagrams relating to this period. Fourth, new types of habitat were created: for the first time since the late-glacial, open sites became available in which grasses, herbs and weeds could flourish. Where sites were cleared by man and then abandoned, scrub or secondary forest replaced the original climax forest; on poorer soils, since the destruction of the forest had inevitably robbed the ecosystem of a major nutrient reservoir, heath began to appear. Fifth, new species were introduced: Neolithic man brought with him, either deliberately or accidentally, primitive cereals, flax and a number of new weeds, such as white campion (*Silene alba*) and charlock (*Sinapis arvensis*).

Not all areas of the country were equally affected. By about 2000 B.C., towards the close of the Neolithic in Britain, a good deal of the chalklands of south-east England had been cleared of forest, but elsewhere the impact was less consistent. The Lake District, Pembrokeshire (Dyfed) and many other coastal areas of western Britain appear to have been settled, but forests on the heavier soil and in the wetter inland parts of the country appear to have remained untouched.

The Bronze Age in Britain (*c.* 1700–500 B.C.) saw a continuation and extension of the same kind of activities already apparent in the Neolithic. The succeeding Iron Age was in some ways more significant in terms of forest destruction. The introduction of the iron axe presented a much more effective way of felling trees; iron was much more widespread than bronze had ever been, and the smelting of iron ore required considerable amounts of charcoal, itself a substantial waster of trees. Moreover, there was also a radical change in farming technique with the introduction of a plough with an iron sherd, replacing the primitive digging stick. This meant plots could be kept in more or less permanent open cultivation.

Thus by the time of the arrival of the Romans, we know from historical accounts that much of lowland Britain had been cleared of forest, except for such heavy areas as the Weald of Kent and Sussex. Highland Britain still possessed a fairly complete cover, not taking on a more modern open aspect until after A.D. 400, when substantial clearances were affected. For the next 1500 years, what was left of areas of forest or waste in Britain continued to be steadily reduced. Open-field systems, enclosures, the creation of estates, the demands on timber for shipbuilding and iron-ore smelting all contributed to this trend. It is only in the twentieth century that any real effort has been made to reverse this pattern and conserve our forest resources and other semi-natural habitats. This is a theme we shall return to in Part Four.

Suggested Further Reading

Cox, C. B., Healey, I. N., and Moore, P. D., *Biogeography* (Chapters 6–8), Blackwell, Oxford, 1973.
Kurten, B., *The Ice Age*, Putnam, New York, 1972.
Pennington, W., *The History of British Vegetation*, E.U.P., London, 1969.
Seddon, B., *Introduction to Biogeography*, Duckworth, London, 1971.
West, R. G., *Pleistocene Geology and Biology* (Chapter 13), Longmans, London, 1968.

CHAPTER TWENTY-THREE

MAJOR NATURAL ECOSYSTEMS

The ecosystem concept can be applied to all spatial levels, from the total biosphere itself to a window box. This chapter is concerned with large areas of the world within which ecosystem structure and functioning are broadly similar throughout.

A fundamental distinction can obviously be made at this global scale between the marine ecosystem on the one hand and terrestrial ecosystem types on the other. The **marine** ecosystem is basically a much larger and more complicated version of the freshwater pond system described earlier (Chapter Nineteen). In both cases, depth of water presents an important control on the forms of life present. However, whereas large plants are common in pond systems, they are almost totally absent from oceans. The structure of the marine ecosystem is very similar for all oceans and it can be considered as one unit, but in the case of terrestrial ecosystems, a threefold division needs to be made.

Forest ecosystems have a complicated layered structure in which the dominant plants are tall and create a variety of ecological niches beneath them. **Grassland** ecosystems have a simpler and more uniform structure, although intermittent trees may occur, as in savanna. In cold or warm **biological deserts**, vegetation is low-growing, and the plants and animals have to tolerate extreme conditions. Despite the link between terrestrial ecosystem types and climate, it is very difficult to ascribe precise climatic limits to them, and for this reason they have not been used as a basis for the identification of climatic zones (Chapter Fifteen). Edaphic and anthropogenic factors are sometimes equally as important as climate in creating these 'natural' systems.

Forest Ecosystems

Forest is the ultimate vegetation type which results from the process of succession on land areas, unless local conditions such as climate, soil or biotic factors arrest development at an earlier seral stage (Chapter Nineteen). When the situation is favourable for the growth of trees, they become the dominant plants of the community for two main reasons. First, trees grow taller than other plants and their canopy establishes a micro-climate which has great influence on the plants and animals living beneath. Second, trees live for a long time and therefore the conditions they impose are maintained throughout the life-span of other organisms. However, although the longevity of trees enables them to outlive everything else, they have great problems in regenerating their own kind. The germination and growth of seedlings is precarious as they often have very narrow tolerance ranges and take a long time to mature: for example, pines take about twenty years and beech trees take as long as forty years.

Once the forest is established, it forms a complex ecosystem with a large standing crop. Much of this is made up of the wood of the trees themselves, which is non-productive but represents energy stored in the system as biomass. The productive part of the trees—the leaves—forms an extensive surface area for photosynthesis, so that a lot of food will be available from the primary level. The amounts of food produced and the paths the energy contained in the food take through the ecosystem, will vary with such factors as the species of the dominant trees and the climate. Generally, however, the food webs are complex, the detrital food chains being responsible for the majority of energy flow taking place. The nutrient cycling patterns will also vary with the ecosystem structure and the climate. The presence of a large standing crop means that many nutrients will be locked within organic material: trees immobilise nutrients for much longer periods than annuals or short-lived perennials.

Trees can be divided into the two main types: **evergreen**, which always have leaves, and **deciduous**, which have no leaves at all at some stage, usually in the winter or dry season. Both evergreen and deciduous trees vary greatly in their ecological tolerances, and so have many habitat possibilities. It is estimated that two thirds of the world's land area is still covered in forest, occupying many different climatic zones; not surprisingly, there are significant contrasts within the ecosystem type. Many naturalists have devised differing classifications of forests in relation to climatic zones, but there is general agreement in distinguishing **boreal, temperate deciduous** and **tropical rain forests**.

Boreal Forest

The boreal forest formation is a vast expanse of coniferous, evergreen forest extending across North America and Eurasia on the southern margins of the tundra zone, in a belt of approximately five hundred miles breadth from north to south (Fig. 23.1a). The area occupied by this formation has been subjected to severe glacial or periglacial activity and has much subdued relief and surface water. The conditions for life are harsh because of the adverse climate. The growing season is only of three or four months' duration and even during this time, the energy input from solar radiation is small because of the high latitude. Temperatures are low throughout the year, although the average temperature of the warmest month of the year is higher than 10° C. In the winter the temperatures fall to many degrees below freezing and permafrost frequently extends into the northern edge of the forest. Precipitation ranges from 400 to 700 mm per annum, mostly falling as snow, the weight of which may cause mechanical damage to the trees. Despite the climate, coniferous trees form a dense canopy which intercepts a great amount of light and precipitation so that conditions beneath are dark and dry. Consequently there is little opportunity for undergrowth to develop and very few other plants are associated with the coniferous trees.

The trees themselves show very little variety across the formation; species of pine, fir and spruce tend to be dominant throughout. As these are all evergreen they have their photosynthetic equipment ready for use as soon as conditions allow it. This partly compensates for the short growing season, but the primary productivity is still low compared to that of other types of forest. This is not only evident from the slow accumulation of biomass when regrowth occurs, but is also apparent in the ecosystem structure. The low primary productivity means that only a limited amount of energy is available for use at the secondary levels, and typically there is a very small amount of animal biomass as well as a lack of diversity. At the herbivore level the invertebrates are predominant, the vertebrate herbivores only becoming numerous in seral areas where foliage is thicker. Carnivores, such as the wolf and lynx, and the large omnivores, such as the black and grizzly bears, which need a lot of food to maintain themselves, are scarce. Studies conducted in the boreal zone indicate that the populations of animals inhabiting them oscillate markedly. Many ecologists interpret this as a sign of instability.

The combination of coniferous dominants which are low in nutrient demand, the lack of diversity, and the climatic conditions, results in slow, impoverished nutrient cycles. Few nutrients will be taken into the plants and few will be returned to the soil in the leaf litter. However, there will be a continual return as the leaves fall from the evergreen trees throughout the year. Most decomposition is fungal since bacterial activity will be slow in these conditions, and the resulting humus is the mor type. Characteristically the boreal forest is found growing on podzols which tend to become highly acidic.

Temperate Deciduous Forest

This type of forest, dominated by broad-leaved deciduous trees, had a great extent in the past when it covered most of the temperate areas of Europe, eastern North America, eastern Asia and small parts of South America and Australia (see Fig. 23.1b). The temperate deciduous forest has probably been more modified by human activity than any other type of ecosystem. Most of it has been destroyed in historic times but we can attempt to reconstruct its main characteristics from evidence provided by pollen analysis and by studying the few remaining areas.

The climatic zone it occupies is less extreme than that of the boreal forest. There is a longer growing season, higher light intensity and a moderate amount of precipitation of between 500 and 1500 mm per annum. The temperature regime is also characterised by lack of extremes but there is still a marked cold season which plants and animals must endure.

The dominant trees are more varied than in the boreal forest: for example, in Europe there are twelve main dominant species including different sorts of oak, beech and ash, and in North America, which is richer in flora, there are at least sixty dominant species, notably several sorts of chestnut, maple and hemlock. The deciduous habit and the lighter shade cast by these trees compared to the conifers, allows sufficient light to reach beneath the canopy so that understorey vegetation can develop. This will vary dramatically with the tree species present. Beech casts quite a dark shade and consequently few plants grow below it, but oak and ash cast light shades and layers of shrubs and herbaceous plants may develop producing a species-rich community. Some undergrowth plants, such as the bluebell, are adapted to use the light available before the leaves grow on the trees in the spring. As soon as the temperature becomes warm enough these plants have a quick spurt of growth, flower and reproduce using the light energy prevailing in the absence of the canopy, then exist vegetatively for the rest of the year.

The primary productivity of the temperate deciduous forest is much greater than that of the boreal forest. There are larger amounts of biomass at the autotroph level and consequently a larger standing crop can be supported at the heterotroph level. The amount and rate of energy flow through the system is much more than in the boreal forest, and this in turn allows the development of diversity. There are more possible habitats and specialised ecological niches. The food chains are complex and long, and the system seems to exhibit more stability in its animal populations than the boreal forest.

Most of the deciduous trees are nutrient-demanding and therefore the nutrient cycling patterns also show marked contrasts with the boreal situation.

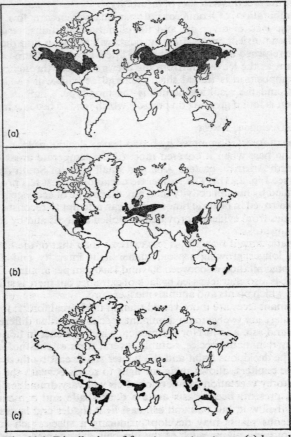

Fig. 23.1. Distribution of forest ecosystem types: (a) boreal forest; (b) temperate deciduous forest; (c) tropical rain forest.
(Reproduced from Cox, Healey and Moore:
*Biogeography: An Ecological and
Evolutionary Approach*, 2nd edition, 1976,
by permission of Blackwell Scientific Publications)

Larger amounts of nutrients are used and their movement is more rapid. There is a bulk return of nutrients from the trees with the leaf fall of autumn. Characteristically the leaf litter is nutrient-rich and decays by the action of bacteria to form mull humus. The soils associated with the temperate deciduous forest are varied but on the whole they are brown earths.

Tropical Rain-forest

The tropical rain-forest occupies low-altitude areas near the equator in South America, Central and West Africa, and in the Indo-Malay peninsula and New Guinea regions (see Fig. 23.1c). Although these areas are physically isolated, the forest growing in them shows great similarity of structure and function. It is a broad-leaved evergreen forest of dense, prolific growth and an extremely diverse fauna and flora. The hot, wet tropical climate is highly conducive to plant growth and there is very little seasonality which means that the growing period extends throughout the year. This, combined with the large energy input to the system from solar radiation in the low-latitude areas, results in high rates of primary productivity, and a large standing crop of vegetation, which can support a great deal of animal biomass. In these conditions there will be severe competition for survival, leading to specialisation of roles and the predominance of narrow ecological niches. All green plants strive to reach the light so that they either become very tall, or adopt a climbing habit or live as **epiphytes** (plants living on other plants but not deriving food from them). Beneath the tree canopy, which may itself consist of two layers, there is usually a well-developed layering of understorey vegetation, which is so dense that hardly any light reaches ground level.

The dominant trees are extremely varied in species but have similar appearances, typically characterised by buttress roots, dark leaves and a thin bark. The leaves possess thick cuticles for protection against the strong sunlight, and drip tips whose probable function is to shed water rapidly, thereby aiding transpiration. The heterotrophs also show similarities in their general characteristics. Many snakes and mammals are adapted to live in the trees because this is where the bulk of the foliage exists.

Providing the tropical rain-forest is undisturbed it is the most diverse and productive type of forest ecosystem, but if the canopy is depleted the soils soon become infertile. Nutrient cycling is rapid, as the vegetation is demanding, and decomposition is accomplished quickly by bacterial action. Very few nutrients are stored in the soil and the system exists in a precarious state of balance which can easily be upset.

Grassland Ecosystems

The grass family (Graminaceae) exhibits a remarkable range of tolerance to habitat factors, and although most species are found in temperate or tropical regions, it has a world-wide distribution. Grasses can be annual, biennial or perennial in habit. Despite a variation in species height from a few centimetres to several metres, all grasses have a similar life form and all are herbaceous (non-woody), with the exception of the bamboos. The principal taxonomic feature of the family is that all species have sheath-like leaves produced alternately from nodes on the stem. Another important characteristic is that grasses grow from their base; consequently they can tolerate grazing or burning better than most plants. They also have extensive root systems, which, coupled with their efficient reproduction and dispersal capabilities, means that once a grass sward has formed, it is very difficult for other species to invade it.

Grassland ecosystems contrast with forest ecosystems in several ways. They have a much smaller biomass, of which a large percentage is made up of roots. In the part above ground, the simple growth form precludes any **obvious** structural layering. A ground layer of small plants with 'rosette' forms, mosses and lichens is one of the more consistent features. Similarly, there is not a great deal of micro-climatic variation within a sward, except in tropical grasslands, where a discontinuous layer of trees or shrubs may diversify the structure. However, a grass sward as a whole usually has a higher albedo than a tree canopy and protects the soil from direct heat. Grasses are probably not as effective at precipitation interception as trees, except for the period of maximum growth. The grass form facilitates stem flow, and surface run-off is greater from grass-covered than from forested slopes.

The annual primary productivity of a grassland ecosystem is only about an eighth or ninth of an adjacent forest area. The smaller standing crop also means that there are more limited nutrient reservoirs in grassland. Turnover of nutrients is relatively rapid. Since much of the plant food occurs in the soil, a large soil fauna, notably made up of decomposers, is a characteristic feature of the ecosystem type. Surface animal species are frequently large: herbivores (bison, antelope, horse) adapt their behaviour to the relatively open and unprotected habitat by congregating in herds and being fleet of foot. Smaller mammals, such as mice, voles and rabbits live below but feed above ground.

The ecological status of grasslands has given rise to much debate. Early workers thought that all extensive grasslands were too dry for trees and therefore represented a climatic climax. More recently, increased study of the fauna and the use of pollen analysis has led to the conclusion that fire and human activity are important elements in these ecosystems, even though we normally refer to them as 'natural'.

The most extensive natural grasslands occur in sub-humid or semi-arid areas which have a low variable rainfall with a marked spring or summerl maximum. They attain their widest extent in continental areas of low relief, forming extensive plainlands. Two main types of grassland are normally distinguished (see Fig. 23.2a): temperate grasslands, in which woody growth is absent or negligible, and tropical grassland (savanna) in which scattered trees are much more common.

Temperate Grasslands

These include the prairies of North America, the steppes of Eurasia, the pampas of South America, and the veldt of South Africa. Smaller tracts occur in Australia and New Zealand. Precipitation in these areas ranges from 250 to 1000 mm per annum, and the grasslands extend over a wide range of soil conditions. Trees only occur on steep slopes or near water. The geographica isolation of these areas from each other has led to some species differentiation, but most other features are similar.

The **prairies** have been the most closely studied. They exhibit a gradual transition of character from east to west, partly related to climate. Areas of so-called 'true' prairie, composed of grass two or three metres high, were originally widespread in the wetter east in Michigan and Illinois, but have now almost completely vanished under extensive cultivation. Remaining remnants merge westward into mixed prairie, about one metre high, which itself is replaced by short-grass prairie on the high plains. Here, grasses are only ten to fifteen centimetres high, and in the lee of the Rockies, become mixed with cactus and sagebrush. Short-grass prairie, formerly attributed solely to the increasing westwards aridity, is now thought to be partly a function of grazing pressure: if this is eliminated, medium grasses reappear and become dominant. Soils within the prairie belt include deep and fertile chernozems, prairie soils and chestnut soils (Chapter Eighteen).

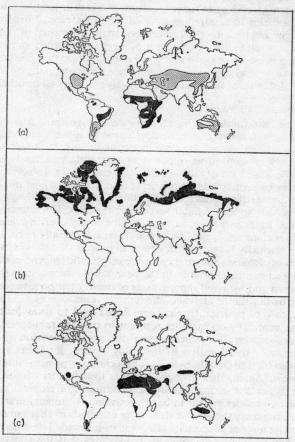

Fig. 23.2. Distribution of (a) temperate grasslands (*light shading*) and tropical savanna (*dark shading*), (b) tundra, and (c) arid areas.
(Reproduced from Cox, Healey and Moore:
*Biogeography: An Ecological and
Evolutionary Approach*, 2nd edition, 1976,
by permission of Blackwell Scientific Publications)

Although the boundaries of the prairies seem to have fluctuated in the geological past, one of the most interesting features is the sharpness of the prairie edge, and the lack of an **ecotone** or transitional area with adjacent forests. The factors which have created this situation are difficult to isolate, but most authorities would include grazing and fire. Large herds of bison once roamed the prairies; it is estimated that in 1600 there were as many as forty-five million, an important force in maintaining the general uniformity of the whole habitat. The American ecologist Carl Sauer particularly favours fire as a necessary condition of prairie maintenance: forest remnants can be shown to exist only where steep slopes form a natural fire-break. Early white settlers noted that the Plains Indian frequently used fire as a means of herding bison. Sauer has suggested that the prairies are essentially a product of post-glacial times, but others relate the development of large herbivorous animals in the Tertiary to the presence of open grassland habitats. One possible conclusion is that the prairies, and other similar grasslands, may originally have been of a climatic origin, but they have been maintained and extended by other factors.

Tropical Grasslands (Savannas)

The savanna lands of Africa, South America and Australia (see Fig. 23.2a) are essentially open, and ecologically dominated by a herbaceous stratum in which grasses and sedges are the principal components. However, the much greater diversity of tropical as opposed to temperate grasslands is often a function of the added variety afforded by wooded plants. In some cases the tree cover may be as much as 50 per cent; in others it may be nil. All types experience a climate of marked seasonal drought, and many of the plants, both grasses and woody species, exhibit xerophytic features. The latosolic soils of savanna areas frequently include near-surface lateritic crusts, creating an impermeable surface soil layer in which nutrients, especially phosphates and nitrates, are markedly lacking. It has been suggested that these edaphic factors, as much as climatic controls, favour the xerophytic tendencies of savanna species. Marked contrasts exist in the appearance of the savanna during the year: the brown and withered short grasses of the dry season give way rapidly to tall lush growth with the arrival of the summer rains.

As in the case of prairies, tropical grasslands tend to show little ecotone development, especially on margins adjacent to tropical rain-forest. Overall, savanna boundaries on all continents reveal only poor correlation with precipitation amounts or the duration of the rainy season. Botanists such as A. Schimper, who tried to correlate savanna as a climax type with climate, found this impossible. Again, this strongly suggests that factors of soil, fire and grazing are important in maintaining the character of tropical grasslands. Many of the tree species appear to be fire-resistant. Although human occupation of the African savanna may extend over hundreds of thousands of years, fire and grazing are not necessarily always man-induced. The great variety of herbivores and carnivores in these areas indicates that the ecosystem is of considerable antiquity. It has been proposed that tropical grasslands have evolved over a long period of time which has been accompanied by progressive soil degradation. If so, savannas are now **edaphic climaxes**, depleted further in many areas by burning and grazing to treeless grasslands. Thus they persist in many areas where the climatic conditions are suitable for tree growth.

Biological Deserts

Very few areas of the world, apart from ice-sheets, are absolute deserts, devoid of any form of life. On the other hand, there are extensive regions where biomass and organic productivity are very low. The largest areas where

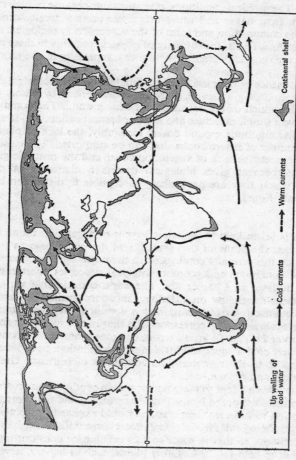

Fig. 23.3. The marine ecosystem: oceanic circulation features and the extent of the continental shelf.

Continental shelf

Warm currents

Cold currents

Up welling of cold water

this is the case are climatically conditioned, either by a lack of water as in hot deserts, or by extreme cold as in tundra regions (polar or cold deserts).

Despite the obvious climatic and geomorphological differences between arid and cold deserts, they have certain basic ecological characteristics in common. In both, the environment is harsh, requiring a high degree of tolerance or adaptation in the organisms present. Plants and animals of biological deserts tend to show less species variety than other ecosystems, but more specialisation. The physical conditions, often unstable because of high rates of wind deflation, frost heave and other processes, exert a greater ecological influence on the composition and form of the vegetation in biological deserts than elsewhere. As a result, the diversity of physical habitats in these regions is reflected in the mosaic of small communities, especially in the more hilly regions.

The virtual absence of trees is a marked visual feature of biological deserts. With the exception of giant cacti, they are characterised by plants of low growth of which much of the organism is below ground. The communities themselves have a simple structure and a lack of stratification, and frequently form only a discontinuous ground cover. Inevitably, the lack of plant food restricts the number of heterotrophs that can be supported. The close inter-relationship between the lack of vegetation cover and the unstable geomorphological environment gives biological deserts a character of delicate instability in which they are particularly susceptible to disruption by man (Chapter Twenty-Four).

Tundra

The tundra is taken here to include all types of vegetation found in high latitudes between the limits of tree growth and the polar ice-caps (see Fig. 23.2b). In effect, this virtually confines us to the northern hemisphere. Some of the ground conditions and geomorphological processes operating in the tundra were outlined in Chapter Six. There is a considerable variation in surface processes depending on lithology, moisture availability and slope gradient, but solifluction over permafrost is common over large areas. The climate of the tundra roughly corresponds to that of the sub-polar zone described in Chapter Fifteen. In broad terms, climates range from a continental type of extremely cold winters and little snow precipitation, as in Siberia and northern Canada, to the raw maritime conditions of southern Greenland, northern Norway and Alaska.

Plants in the tundra adapt to these unfavourable conditions in several ways. The main problems are posed by the low temperatures and the short growing season. Adaption to sub-zero temperatures would appear to be both physiological and morphological. Physiological protection entails changes in the cell sap and protoplasm, so that in some species cellular ice does not form until temperatures drop below −30°C. Other plants, such as lichens, never freeze and can adjust to rapid and extreme temperature changes. The low growth form of tundra plants represents a morphological adaptation which takes advantage of slightly higher temperatures near the ground and avoids extreme wind exposure.

The shortness of the growing season—less than three months—is to some extent offset by long day length in summer. Nevertheless, the growing season poses major problems for productivity and reproduction, and plants must complete their development cycle in the time available. Photosynthesis in most tundra plants starts immediately soil temperatures rise above 0°C, particularly in evergreen shrubs. Deciduous plants may possess partly developed leaves, formed in the previous autumn, which allow the plant to take advantage of the whole of the new growing season. Some reproduction in tundra plants takes place vegetatively, some by seed production, which may need two or three seasons to be completed. Flowers, often vivid, are formed in one season, and seed maturation occurs in the next. In still other plants, germination commences before the seed is dispersed. However, not surprisingly, annuals are rare.

The importance of these limiting factors means that the principal regional variations in tundra vegetation are related to the northward reduction of the duration and warmth of the growing season, combined with increasing winter severity. Where undisturbed, the southern parts of the tundra are characterised by stands of dwarf willow, birch and alder, sometimes up to two metres in height. Further north, these give way to heaths of cowberry or crowberry or to intermittent swards dominated by rushes and sedges over peat. Under the severest conditions, all plants but mosses and lichens are eliminated, and large areas of bare rubble or rock are to be found. Superimposed on this broad framework are communities of vegetation dependent on local conditions. In particular, relief variations affect the depth and duration of snow cover, as well as soil mobility, and these factors in turn produce local vegetation sequences, or **catenas**, related to altitude and slope angle.

Animals in the tundra are limited in number and variety by the lack of plant food and the intense cold. The difficulty of finding shelter is compounded by the impossibility of underground refuge where permafrost exists. Warm-blooded animals must either be protected from the surface cold by such adaptions as woolly coats and low surface area to body volume ratios, as in the polar bear, or they must migrate. Cold-blooded animals, of which the insects are by far the most numerous, can survive in larval form throughout the winter. The main herbivores include caribou, reindeer, musk ox, lemmings and the Arctic hare; predators, both carnivores and omnivores, include the Arctic fox, the wolf and bears. The total animal biomass is relatively small and undergoes marked seasonal fluctuations in volume. The lack of animal diversity is reflected in the small number of trophic levels, although food webs are often made complex by the relatively high proportion of omnivores in the system.

Some authorities are of the opinion that the simplicity of the tundra ecosystem renders it inherently unstable. One possible manifestation of this instability is the great periodic fluctuations that occur in the population of such species as lemming or Arctic fox. Because the types of prey at any particular trophic level are so few, any variation in numbers has major repercussions on those of the next higher level. It is difficult to decide whether the lack of diversity in the ecosystem is entirely a result of the rigour of the climatic conditions, or whether it also reflects the shortage of time in which the system has been able to develop. The tundra in its present geographical position can only date from the recession of the ice-sheets after the last glaciation, although tundra as an ecosystem type may have first appeared in the late Tertiary.

Arid Areas

Up to a third of the world's land surface can be described as arid (see Fig. 23.2c), characterised from a biological point of view by a lack of water availability rather than a complete absence of water. The ecosystem type described here refers mainly to the hot arid zones of the world, such as the Sahara and Australian deserts. Cooler deserts—for example, those of the Gobi and Patagonia—have been relatively little studied, but possess some of the limiting factors of both the hot deserts and the tundra. Hot deserts occur in the subtropical dry zone of the global atmospheric circulation system (Chapter Fifteen). The majority of accounts of the plant and animal life of these areas have been largely descriptive; much less is known about ecosystem functioning.

With one or two notable exceptions, such as the creosote bush (south-west USA), desert plants capable of withstanding extreme desiccation of their tissues or very high temperatures are in fact relatively few. Up to 60 per cent of desert floras are made up of annual or ephemeral species which evade the drought by completing their life cycles within a few weeks of the onset of any rain, their seeds remaining quiescent throughout the succeeding dry period. Perennials, on the other hand, are faced with the year-round conflicting problem of avoiding desiccation and keeping cool at the same time. One of the most important ways of avoiding water loss is to close the leaf stomata, particularly during the hottest period of the day, yet the stomata need to be kept at least partly open to maintain transpiration and cool the leaves. Some plants only open their stomata at night. Succulents, such as cacti, combat the water problem with the aid of well-developed water storage organs and small surface area to volume ratios.

In addition to the problems of physical drought and great heat, many arid zone plants have to tolerate the physiological drought set up by saline ground conditions. An evasion of the worst effects of high salt concentrations may be achieved by the synchronisation of life cycles with rain periods sufficiently wet to leach temporarily the upper soil layers. Thus besides their xerophytic characteristics, many plants need also to be halophytic.

The most noticeable visual characteristic of areas of desert vegetation is the discontinuous cover and the even spacing of individuals. This appears to be the result of extensive root development and competition. Vegetation is more discontinuous than in the tundra, but on the other hand it is more diverse in composition and form. Floras of discrete areas have tended to evolve in isolation from each other, whereas the tundra is more continuous. Related to varying degrees of aridity, desert vegetation includes low woody scrub formations, cacti communities, intermittent swards of perennial grass tufts, ephemeral or seasonal herbaceous vegetation, and 'accidental' vegetation in areas where rainfall may occur only once in several years. Under the most rigorous conditions a combination of a lack of precipitation, intensity of evaporation and high soil salinity or mobility (e.g. on dunes), may exclude vegetation entirely. These absolute biological deserts are, however, localised and limited in extent. In the particularly harsh conditions of the coastal deserts (see page 179), where the sole source of moisture is the sea mist, only halophytes or succulent epiphytes, absorbing moisture directly from the atmosphere, can survive.

As in the tundra, the animal species of deserts are fewer but more specialised than in humid environments. Protection against water loss and high levels of body heat pose similar problems as for plants. Morphological protection may be given by such features as an impermeable body covering, a small number of sweat glands and a light colouring. Camels and donkeys have in addition a physiological tolerance of high water losses and can survive a water reduction equal to more than 25 per cent of body weight. In smaller animals the burrowing habit is widely developed, especially among insects. The atmosphere of the deeper soil layers remains humid, and provides a more temperate climate than the soil surface or the free air above it. Nocturnal activity and summer dormancy are also common features.

The sparseness and marked seasonal variations of biomass at the autotroph level is reflected in the widely fluctuating populations of some of the animal species in both time and space. Synchronisation of breeding cycles with periods favourable to vegetation growth enhances the explosion of life that occurs in deserts within a few days of the onset of rain. A detrimental side-effect of this synchronisation occurs where man irrigates desert areas and provides an abundant source of food without predators. Spectacular locust plagues are one result of this. Desert ecosystems are as precariously balanced as that of the tundra.

Archaeological evidence suggests that desert margins have long been useful to man, either for grazing or, with irrigation, for crops. Irrigation tends to raise the water-table and can bring salts to the surface, ultimately rendering the ground sterile. This, coupled with widespread grazing, has created considerable areas of man-made desert. Although the desert ecosystem type can be said to have evolved because of climatic control, hardly anywhere today are its margins 'natural'.

The Marine Ecosystem

The basic functional mechanisms common to all ecosystems (Chapter Nineteen) apply as much to the sea as to the land, but major differences exist between the environmental conditions and life-forms in the marine ecosystem and those of the terrestrial types. Oceans cover 70 per cent of the surface area of the world, they are habitable throughout and support a total biomass probably as much as ten times that on land. In many ways, the marine environment is much more favourable to life than land areas; it is more equable, and the two most essential gases for life, oxygen and carbon dioxide, are readily available in water, provided it is not polluted. In addition, many of the nutrient minerals found in the Earth's crust are dissolved in the sea in varying amounts.

Environmental Conditions

The main environmental gradients in the sea are related to temperature, salinity, and light intensity. Salinity is caused by at least forty-five elements, the major two being sodium and chlorine. The most saline conditions occur where temperatures, and hence evaporation, are highest; the Red Sea has an average value of forty grams of salt per thousand of sea water (40‰). The lowest values occur near melting ice or near river mouths; the Baltic Sea has a salinity in summer months of barely 5‰. In the open water of the major oceans, the range is much less, from 37‰ in the tropics to 33‰ in polar seas. Many marine organisms have very narrow tolerance ranges to particular salinity concentrations, which may therefore localise them considerably in terms of depth or area.

Temperature variations in the sea are much less than those on land. The difference between the surface temperature of the warmest seas (32°C) and the coldest (−2°C) gives a range far less than that of land (about 90°C). Around the coasts of the British Isles, surface water temperatures vary annually from about 8°C to 17°C. Both vertical and horizontal ocean currents (Fig. 23.3) play a major role in equalising variations of temperature, salinity and dissolved gases in the oceans, as well as being important factors in the global energy budget.

The availability of light exercises as much fundamental control on the basic process of photosynthesis in the sea as it does on the land. The amount of light reaching the surface varies with latitude and with season; much is lost by reflection from the water surface in high latitudes and when the sea is rough. Absorption by the water increases very rapidly with depth; this rate is also affected by the amount of turbulence and suspended matter in the sea. The depth at which the compensation point occurs, i.e. the point at which the light is not sufficient to allow photosynthesis to proceed at a rate which compensates for the rate of respiration, varies considerably; it may be as little as ten metres in inshore waters, as much as a hundred metres in open water. The latter figure approximately separates the **euphotic** zone, in which there is enough light for photosynthesis, from the deeper **disphotic** zone lying between one and two hundred metres below the surface. Light penetrates into the latter zone but it is too dim for photosynthesis. It is noteworthy that the maximum depth of the euphotic zone roughly coincides with the average depth of the continental shelf.

Plant Life: Primary Production

Marine plants are confined to the euphotic zone by the light factor. They are far less diverse than land plants, being dominated by algae, with only a few angiosperms present, most of which are found in the near-shore zone. The most obvious and visible types of marine algae are seaweeds, but about 99 per cent of marine vegetation is made up of microscopic floating algae (**phytoplankton**). These are one-celled organisms containing chlorophyll, and include diatoms and dinoflagellates. The phytoplankton are responsible for all but a minute fraction of the primary productivity of the marine ecosystem. At this level total productivity is probably less than on land, mainly because of the relative lack of nutrients in the euphotic zone. Production varies with time and place in the ocean; this is related to light, temperature and particularly to the availability of nitrate and phosphate nutrients. Nutrients absorbed by phytoplankton sink to deeper layers of the sea when the plants, or more usually the animals which feed on them, die. The return of the nutrients for re-use is dependent on upward-moving sea-water. Thus maximum phytoplankton production occurs in areas of upwelling sea-water, as off the coast of Peru (see Fig. 23.3), or where turbulence and mixing of water is initiated by waves, as on continental shelves. Near-shore areas additionally receive nutrients from rivers. Coastal and estuarine areas therefore have a high productivity and great diversity of plant life, making them among the most fertile parts of the marine ecosystem.

In cold and cool temperate seas, marked seasonal variations in productivity occur, low phytoplankton activity being characteristic of winter days. Bursts of production occur in spring and on a lesser scale in autumn, coinciding with more favourable light and temperature conditions and equinoxial gales. In summer, calmer conditions may allow a **thermocline** to develop, a sharp temperature inversion between warm surface waters and colder waters lower down, which effectively prevents water interchange and hence the return of nutrients to the euphotic zone. In the tropics, although light and temperature conditions are such that plant growth can continue throughout the year, average annual production is low because the thermocline tends to be a more permanent feature, slowing nutrient replenishment. The open waters of the tropical doldrum zone (Chapter Fifteen) are to some extent comparable in this respect to biological deserts on land.

Animal Life: Food Chains

Marine fauna is very diverse, and food chains tend to be long and complex. Because of the tiny size of phytoplankton, large herbivores such as those found on land feeding directly on the bulkier land plants, do not exist in the sea. The bulk of the grazers of phytoplankton are the **zooplankton**, minute marine animals who convert the plant food into more manageable proportions for slightly larger animals, and so on. Most of the phytoplankton appears to be cropped live, relatively little of it entering a detrital food chain. This is a marked contrast to the terrestrial situation, where a considerable proportion of dead plant matter enters the soil to support saprophytic organisms.

The marine zooplankton are generally larger in size than the phytoplankton and are more diverse in form, ranging from one-celled animals to young jellyfish. Among the more important of a great number of species are such small crustaceans as the copepods, one of the major sources of fish food in the northern hemisphere, and the shrimp-like 'krill' population, which provides the main diet of Antarctic whalebone whales. Although the distribution and numbers of zooplankton depend on the available phytoplankton, they are not directly dependent on light, and can exist to all depths, probably deriving food from detrital material which sinks down. Apart from zooplankton, the only other direct plant feeders are some of the bottom-living (**demersal**) fauna of shallow seas. In deeper waters, the zooplankton forms virtually the only link between the primary producers and other forms of marine life.

Zooplankton is fed on directly by many surface water (**pelagic**) fish, including mackerel, sardines and herring (see Fig. 19.2) in cooler waters and basking shark and tunny in warmer waters. It also provides food for the great variety of marine invertebrates (e.g. molluscs, worms, prawns), which are in turn eaten by bottom-dwelling carnivores such as haddock, cod and plaice. However, at this level, food chains become complicated; there are carnivorous invertebrate predators or scavengers which may prey on or compete with demersal fish for food. In addition, certain demersal and pelagic fish prey on each other. At the top of the marine food chain, the fish themselves are eaten by sea-birds, seals and man.

Man's activities as part of the marine ecosystem have been confined largely to the uppermost trophic level. Exploitation of marine life has been almost solely concerned with animals large enough in size and numbers to make them worth catching. Only the seaweeds provide a direct, albeit infinitesimal, link between the primary production and man. Increasingly, however, primary production is being affected indirectly by the pollution of inshore waters. Because of the complexity of the marine ecosystem, the visible affects of pollution are often only seen at a late stage, when it has already affected all trophic levels. This forms but one example of man's inadvertent impact on natural systems, a theme which will be considered in the next chapter.

Suggested Further Reading

Cloudsley-Thompson, J. L., and Chadwick, M. J., *Life in Deserts*, Foulis & Co., London, 1964.

Eyre, S. R., *Vegetation and Soils: A World Picture* (2nd edn.), Edward Arnold, London, 1968.

Hopkins, B., *Forest and Savanna*, Heinemann, London, 1965.

Isaacs, J. D., 'The nature of oceanic life', *Scientific American*, **221**, 1969, 147–162.

Longman, K. A., and Jenik, J., *Tropical Forest and its Environment*, Longman, London, 1974.

Moore, H. I., *Grass and Grasslands*, New Naturalist Series, Collins, London, 1966.

Richards, P., *The Tropical Rain Forest*, Cambridge University Press, London and Cambridge, 1952.

Schultz, A. M., 'A study of an ecosystem; the Arctic Tundra', in *The ecosystem concept in resource management*, edited by G. Van Dyne, Academic Press, London, 1969.

Tivy, J., *Biogeography*, Oliver & Boyd, Edinburgh, 1971.

PART FOUR
PHYSICAL GEOGRAPHY AND MAN

CHAPTER TWENTY-FOUR

HUMAN IMPACT ON THE NATURAL ENVIRONMENT

Man's relation with his natural environment is a complex one. While he is subject to certain natural controls and events, he also acts as the dominant force in many of the Earth's physical and biological systems. The relationship has changed with time. For thousands of years, the direction and extent of his progress were to a considerable measure dictated by his physical environment, which sometimes presented him with very difficult obstacles. Increasingly, man has become capable of altering his physical environment to suit himself. Although the object of these alterations was to improve his living conditions, in some cases they have created major long-term problems, and in still others they have been catastrophic, both for the natural environment and himself.

This book has attempted to explain the operation of natural systems and processes, but equally it has been stressed at several points that man is an important influence on many of them. In some parts of the world, the environment has been so transformed that few elements of its original nature are detectable. Even extreme habitats such as the tundra or hot deserts only sparsely populated by man have not escaped untouched, since they are often the most sensitive to the slightest interference. Many apparently natural systems are in fact **control systems** in which man acts as a regulator either consciously or inadvertently. At best, except for large-scale weather phenomena, natural systems are mostly **modified systems**. In this chapter we shall consider some of the ways in which climate, landforms, soils and ecosystems have been inadvertently altered by man.

Modification of Landforms

Mining and quarrying, deforestation, the introduction of exotic plants and animals, the use of agricultural machinery, the building and use of tracks and roads, and the overgrazing of pastures, have all, singly and in combination, profoundly altered landforms and caused **accelerated erosion and deposition** to occur. Where man excavates or piles up material himself, he can be regarded as a direct agent of change; where he causes natural landform processes, such as wind and water action, to accelerate or diminish, he is acting in an indirect manner. Indirect effects are by far the most widespread. Much of this influence occurs accidentally or secondarily to some other purpose; conscious attempts to influence landform processes—for example, by building coastal groynes or by reafforestation—are inevitably expensive and limited in extent.

Direct Alteration of Landforms

Man has a direct effect on the shape of landforms by excavating and piling up earth, reclaiming land from the sea and causing subsidence through mining.

These activities have greatly increased since the Industrial Revolution with the development of enormous machine power and explosives for moving material. Railway and motorway construction provides many familiar examples of man-created slopes, embankments and cuttings. **Land scarification** is sometimes used as a general term for disturbances created by the extraction of mineral resources; open-pit mines, quarries, sand and gravel pits are among the forms of scarification. Strip-mining is one of the most devastating examples of landform alteration of this kind. Although common in the United States, it does not occur on a widespread scale in Britain, except as a method of mining Jurassic ironstone in Northamptonshire. The effects of **subsidence** are common in most of the older coal-mining areas of Britain. Switchback roads, perched canals, fractured buildings and flooded depressions or **flashes** are all visible manifestations of recent changes in the surface form of the ground.

Equally obvious as man-created landforms are coal tips and other waste heaps from mining and quarrying. Many of these features are geomorphologically unstable, allowing various forms of mass movement to generate. When saturated by heavy rain, spoil tips are frequently subject to sliding and flowage, supplying sediment that clogs stream channels. In 1966 at Aberfan in Wales, a major disaster occurred on a spring-saturated coal waste heap which moved as an earthflow, destroying part of the village below, including a school and many of its children. Similar problems may arise on other constructed slopes: the large number of earthflows triggered during the building of the Panama Canal is a well-known example. More recently, the building of new trunk roads and motorways in Britain has encountered slope failure in several instances: at Port Talbot, Keele and Sevenoaks, excavation reactivated slope shear planes which were last active under periglacial conditions during the Devensian glaciation. These sites required extensive engineering works to stabilise or avoid the slopes.

Indirect Effects: Slopes and Rivers

By far the most important of all man's effects on landforms are those connected with his interference with the natural vegetation, in particular with the clearing of forest for agricultural purposes. There is a close relationship between the amount of vegetation cover and erosion rates on hillslopes, and hence with the amount of sediment in streams. A stable vegetation cover acts as an effective regulator of natural erosion, protecting the ground from direct raindrop impact, absorbing some of the run-off, and making the slope more cohesive. With the removal of the vegetation, the surface loses its plant litter, causing a loss of soil structure, cohesion and porosity. Overgrazing has similar effects, and the introduction of animal pests such as the rabbit into Australia has also had a detrimental effect on slope stability. Some idea of the effect of vegetation on erosion can be gleaned from Table 24.1, which relates sediment yield to various categories of land use within a small area. The contrast between open cultivated land and forested areas is readily apparent.

Multiple shoe-string rills and gullies on hillsides are often a typical manifestation of man's indirect effect on slopes. They are presently found in many parts of the world, notably in semi-arid regions susceptible to tropical downpours. In an area such as South Australia, the recent date of a great deal of gully and sheet erosion on slopes is testified by the burial of fence-posts and

Table 24.1. Run-off and Sediment Yield from Various Types of Surface (Northern Mississippi)

Land use type	Average annual rainfall (cm)	Average annual run-off (cm)	Average annual sediment yield (tonnes per hectare)
Open Land:			
Cultivated	132	40	50
Pasture	129	38	36
Forest Land:			
Abandoned fields	129	18	0·29
Depleted hardwoods	129	13	0·22
Pine plantations	137	2·5	0·045

Source: Modified from: S. J. Ursic, 1965. US Dept. of Agriculture, *Misc. Pub. No. 970*, p. 49 (US Gov. Printing Office, Washington, DC).

other man-made debris. There is evidence that in some long-settled areas of the world, like western Europe, where gullies are not now a prevalent feature of the landscape, they were more widespread in past times when the natural vegetation cover was first removed. We may note that it is not always easy to distinguish between the effects of man and a changing climate on hillslope erosion. For example, in the Mediterranean area during the latter part of the Roman period, there was an increasing loss of soil fertility, hillslopes became eroded and valley bottoms were heavily silted. This may have been the result of a tendency towards greater aridity, but many experts believe that human overpopulation and overgrazing by goats were important contributory factors.

High rates of hillslope erosion by overland flow are a natural state in some localities, creating **badlands** as in South Dakota where much of the area is underlain by almost impervious clay formations. But badlands can be artificially produced where accelerated erosion proceeds unchecked. Instances of this extreme form of slope degradation became common in the poorly farmed areas of the southern Appalachians in the United States in the 1920s and did much to bring the whole problem of soil erosion and soil conservation to public attention.

The alteration of infiltration and run-off on slopes by modifying the vegetation inevitably has a profound effect on adjacent rivers in at least two respects: by increasing both the discharge and also the sediment supply. There seems little doubt that many of the floods in mid-latitude rivers would not occur if the vegetation in the drainage basin were in its natural state. Evidence has been put forward that the increase in the frequency of floods in recent decades in the river Severn at Shrewsbury is related to improvements in land drainage in the catchment area. The effect of this is to prevent the fields from retaining a large proportion of the normal precipitation and from releasing it slowly. At times of flood, discharge levels become higher and achieve their maximum more quickly. Another way in which discharge levels may be affected in similar fashion is through urbanisation; the ground surface is rendered impervious by buildings, paths and roads, and precipitation is channelled directly to rivers through drains and sewers. Fig. 24.1 illustrates

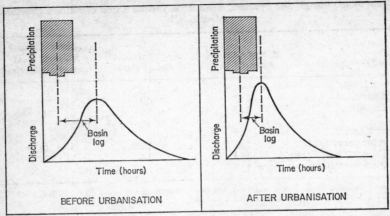

Fig. 24.1. Hydrographs showing the effect of urbanisation on flood lag time and peak discharge.

the effect of urbanisation on flood peaks, with attendant damage to river banks, properties and farmland.

It is difficult to assess quantitatively the importance of man-induced slope erosion on the sediment load of major rivers since we have few pre-interference data. Most investigators are generally agreed that cultivation has greatly increased the sediment load of rivers of south-east Asia, Europe and North America, perhaps by a factor of two or three above the geological norm for the world. The effect of this increase is of considerable importance in the construction of dams and canals; in severe cases, large amounts of sediment supply may also cause valley aggradation, destroying productive land capacity. Specific operations of man which lead to local cases of river silting and aggradation include mining operations, urbanisation and highway construction: all these are sources of excessive sediment. However, much can be done to reduce the effects of building operations—for example, by keeping to a minimum the periods during which bare areas are exposed, and by using sediment basins to trap coarser sediment.

Wind Deflation

The phenomenon of the **dustbowl** in the Great Plains region of America in the 1930s is a well-known example of man-induced land erosion. The area was former grassland underlain by rich brown and chestnut soils, but both overgrazing and ploughing contributed to the catastrophe which caused the widespread abandonment of farms. A great expansion in wheat cultivation in the early years of the decade was followed by a series of droughts; the soil, largely exhausted of its natural fertility, was subject to deflation and particle drifting of disastrous proportions.

The dustbowl situation is by no means unique. In the marginal areas around today's hot deserts, such as the Thar desert of Pakistan and India, and the Egyptian desert, a great deal of deflation is initiated by grazing animals. In other deserts, as in the central Sahara and the south-west United States,

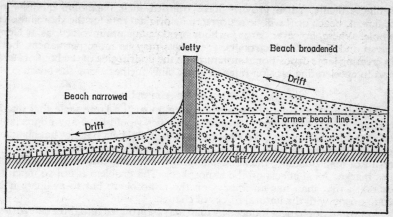

Fig. 24.2. Effect of a groyne or jetty on longshore drift.

desert pavements (Chapter Eight) normally contribute little coarse dust, but this protective layer is easily destroyed by wheeled vehicles, exposing finer-textured materials.

In Britain, coastal dunes are highly susceptible to deflation when interfered with by man. Constant trampling or vehicular traffic quickly destroys the protective grass vegetation, initiating blowouts or landward migration of the dunes. On the Dutch coast, protection of the dune systems from degradation by man is vital as these give protection to large inland areas lying below sea-level.

Coastal Erosion and Deposition

Man can have relatively little impact on the forces that govern waves, tides and currents, but he has had some effect on coastal erosion and deposition at the shoreline by building various structures and by removing beach material for ballast or construction. Before the nineteenth century, the erection of small piers or breakwaters to protect harbours was one of the few ways in which coastal processes could be locally modified. In the last two centuries, the urbanisation of many coastal areas in Britain has often paid little attention to local erosion factors, and one can find many examples where even a few yards of erosion by the sea may spell disaster for a heavily built-up area. Hence various engineering structures such as groynes, breakwaters and sea-walls have had to be built to check marine erosion. However, these are not only extremely expensive to build and maintain, but often defeat the object of the exercise, since by checking erosion in one place they may lead to its increase elsewhere.

This may be illustrated by considering the effect of a single groyne or jetty in checking the movement of beach material (Fig. 24.2). Groynes have been widely used on the shingle beaches of the south coast of England. By preventing the longshore movement of the shingle, this may starve the shore on the downdrift side of the groyne of its supply of beach material, leading to a narrowing of the beach and an increase in direct wave attack on the cliff behind. Ideally, when groynes have trapped their maximum quantity of sediment, beach drift will be restored to its original rate for the shoreline as a whole. Where, however, large harbour breakwaters are involved, as at Newhaven and Folkestone harbours, the effects may be more permanent, both in creating large depositional structures on the updrift side of the breakwaters, and in accelerating erosion rates in chalk cliffs further along the coast.

Modification of the Atmosphere

Atmospheric circulation systems operate on such a large scale that one is perhaps inclined to doubt that man's activities would have any appreciable effect on them. However, it is known that the global heat balance has changed over the last few decades, and we might ask ourselves how much of this is a result of man polluting the atmosphere. It is certainly evident that pollution has marked local effects on the atmosphere. The problem is not so much to establish that man has an impact on the atmosphere but to evaluate it in comparison with the natural forces of change.

Atmospheric changes induced by man may be grouped into three categories: the introduction of solids and gases not normally found in the atmosphere (pollutants); changes in proportions of the natural component gases of the atmosphere; and alterations of the Earth's surface in such a way as to affect the atmosphere. A fourth type of impact, planned weather modification, is considered in the next chapter.

Pollutants in the Atmosphere

To city-dwellers the most obvious way in which man has affected the atmosphere is through pollution. Pollutants include particulate matter, both solid and liquid particles, and gaseous substances such as sulphur dioxide (SO_2), oxides of nitrogen (NO, NO_2, NO_3), carbon monoxide (CO) and hydrocarbon compounds. But not all man-made pollution comes from cities. Isolated industrial activities frequently create a footprint of atmospheric pollution in areas of countryside downwind from the industrial site: particularly infamous examples in Britain include smelters and brickworks. Mining and quarrying activities also send large amounts of mineral dust into the air. Even man-induced forest and grass fires as well as bonfires, can greatly add to particulate pollution at certain times of year.

Atmospheric pollutants are conducted upward from the emission sources by rising air currents as part of the normal convective processes. Larger particles settle under gravity and return to the ground as **fallout**. Smaller suspended particles are brought to the Earth by precipitation as **washout**. By a combination of the two processes the atmosphere tends to be cleaned of pollutants, and in the long run a balance is achieved between the input and output of pollutants, although there are large fluctuations in the quantities stored in the air at a given time. Pollutants are also eliminated from the air over their source areas by winds which disperse the particles into large volumes of clean air in the downwind direction. Smoke stacks are intended to take as much advantage of this as possible. The passage of a cold front accompanied by strong winds is usually very effective in sweeping away pollutants from an urban area, but during stagnant anticyclonic conditions concentrations may rise to high values, sometimes producing a smog (Chapter Sixteen).

Once in the atmosphere, the primary pollutants undergo a number of chemical reactions, generating a secondary group of pollutants. For example, sulphur dioxide (SO_2) combines with oxygen and suspended water droplets to produce sulphuric acid. This acid is harmful to organic tissues and is also very corrosive. **Photochemical** reactions are brought about by the action of sunlight: for example, sunlight acting on nitrogen oxides and organic compounds produces ozone (O_3). Another toxic chemical produced by photochemical action is ethylene.

The harmful effects of atmospheric pollution on plant and animal life are manifold. For humans, many pollutants are irritant to the eyes and dangerous to the respiratory system. During the London smog of December 1962, more than 4,000 additional deaths occurred. Ozone in urban smog has a severe effect on plant tissues; atmospheric sulphuric acid has wiped out lichen growth in many urban areas of Britain. Lead and other toxic metal particles are a particular cause of concern for human health. In addition, pollution also causes many millions of pounds worth of damage to materials: limestone structures suffer greatly in certain British cities unless treated with preservatives.

The global effects of foreign particles in the atmosphere in altering radiation and heat balances is difficult to assess. There have always been major natural sources of particles in the atmosphere, including forest fires and large volcanic explosions. Careful monitoring by U.S. scientists of temperature trends and dust amounts at various heights in the atmosphere has led to the tentative conclusion that man's contribution to atmospheric particles may have far-reaching effects on tropospheric processes—for example, on the rain-making mechanism—but perhaps little effect on processes in the stratosphere.

Changes in Atmospheric Gas Levels

Of the main natural constituent gases in the atmosphere (see Table 10.1), carbon dioxide (CO_2) and oxygen (O_2) are the most critical from an environmental viewpoint, for both are inextricably involved in the biochemical cycles between atmosphere and the surface of the Earth. Although nitrogen comprises four fifths of the atmosphere, its inert chemical nature relegates it to a minor role in this respect. Oxygen and carbon dioxide are naturally added to the atmosphere by 'outgassing' from the Earth's interior. The work of plants has been essential in removing carbon dioxide from the atmosphere and storing it as coal and other fossil organic substances. Before the Industrial Revolution, carbon dioxide levels appear to have been about 290 parts per million (p.p.m.) in the atmosphere. But in the last hundred years or so, this amount has increased by about ten per cent, largely because of man's use of fossil fuels (Fig. 24.3). It has been suggested that, in contrast to the effect of solid particles, an increased level in carbon dioxide content will increase the temperature of the atmosphere, since the gas is an absorber of long-wave radiation. However, although the use of fossil fuels continues to accelerate, we know that temperatures have fallen in the last two decades (Chapter Seventeen) and the link remains unproven.

It has been pointed out also that man's large-scale combustion of hydrocarbon fuels requires a large quantity of oxygen to be withdrawn from the atmosphere and converted into carbon dioxide and water vapour. There is therefore the possibility of a lowering of the oxygen content of the atmosphere to levels which might have a detrimental effect on animal life.

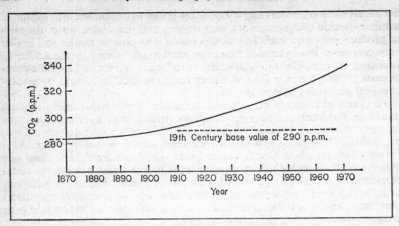

Fig. 24.3. Carbon dioxide content of free air in the north Atlantic since 1870.
(From H. Fløhn, *Climate and Weather*, Weidenfeld and Nicolson)

Changes in **water vapour levels** brought about by man through combustion and alterations to the vegetation cover could in theory markedly affect global radiation and heat balances in the same manner as changes in carbon dioxide levels. But water vapour content varies greatly from place to place and it is difficult to measure global changes. It seems unlikely that there would be a general build-up of excess atmosphere water vapour through combustion, as it would rapidly return to the oceans as precipitation. A special case, however, is the emission of water vapour and various other substances by jet aircraft. These emissions occur in the stratosphere, where the water vapour content is normally very small. The **condensation trails** (contrails) of aircraft can often be observed to spread laterally and develop into cirrus clouds. These clouds are highly reflective and can have an effect on the Earth's albedo.

Alterations to the Earth's Surface

Meteorological processes close to the ground are extremely sensitive to the character of the Earth's surface, and man's alteration of this through deforestation, agricultural practice and urbanisation has had several important effects. One result of these activities is to alter the rate of evapotranspiration. The complete removal of a forest cover will sharply reduce transpiration and thus the amount of water returning to the atmosphere in vapour form. The draining of a swamp will have a similar effect. Just what impact, if any, this has on air masses of large vertical extent, is uncertain.

Another important consequence of surface change is to alter the temperature characteristics of the atmosphere nearest the ground. We noted one example of this in Chapter Sixteen: closely built urban areas develop their own heat island on calm nights in summer. Equally impressive are the changes in the heat budget brought about when an irrigated area is created in an arid region. The albedo of a light-coloured desert area is about 25–30 per cent; there is very little water for evaporation so that all the incoming radiation is available for the direct heating of the air. With irrigation, the albedo drops to

10–15 per cent, and the incoming energy is used up almost entirely in evaporating water. Thus direct heating of the air above the irrigated area is very slight, and day-time temperatures become markedly lower than in the surrounding desert.

A third climatic element that may be modified when man alters the ground surface is the wind. Trees and hedges effectively brake the wind, causing a simultaneous diminution in evaporation and in the carbon dioxide exchange close to the ground. Garden walls or thick tree belts may so effectively still the air immediately to leeward as to cause frost pockets on cold nights.

Modification of Ecosystems

In the latter part of Chapter Twenty-Two we observed that primitive man managed to live as part of the natural ecosystem without altering its main characteristics. Large-scale burning of grasslands may perhaps be regarded as an exception to this. With the beginnings of agriculture, far-reaching effects, both obvious and subtle, were introduced into ecosystems. Man gradually became more sophisticated in knowing just how much to modify an ecosystem in order to harvest the crop he wanted. In achieving this end, he has inevitably simplified ecosystems, disrupted nutrient cycling, introduced alien species and eliminated others, and caused pollution. Only in recent years has there been an awareness of some of the consequences of ecosystem modification. Some of the efforts to redress the balance are considered in the next chapter; this section summarises the more serious consequences of man's impact.

Simplification

The most general effect of man on ecosystems is that he tends to simplify them. This comes about because man's prime concern is to direct energy and material cycling in the system towards himself so that he can easily crop them. Species other than the ones he wants to crop are regarded as weeds or pests, and he attempts to eliminate them. Hence, reduction in species diversity, often to a single species population, is a notable characteristic of man's impact on ecosystems. Food webs are also made much simpler in this process. In arable farming, man removes all other consumer organisms and crops the primary producers. In pastoral farming, he retains a single herbivore species, sheep or cattle, and himself occupies the position of sole carnivore. Again, this brings man into a state of active competition with all plants and animals apart from his favoured ones.

The degree of simplification varies enormously. In remote areas still only inhabited by hunters and gatherers, man may in fact add another trophic level to the rest of the food web. Primitive shifting agriculture in tropical rainforests represents only a temporary simplification and cropping of the natural system as the plot is only cultivated for a few years and then abandoned. On the other hand, grazing economies exhibit a much greater degree of ecosystem simplification. Unless the pastoralism is sufficiently wide-ranging to allow pastures to recover, selective grazing by the domesticated species leads to the eradication of the most palatable plant species, allowing tougher grasses or xerophytic plants to predominate, thus simplifying and downgrading the pasture. Modern arable farming represents perhaps the most extreme form of simplification, producing a highly artificial type of ecosystem.

Ecosystem simplification of this type often results in disastrous side-effects.

A single species population, such as a field of wheat or a herd of cows, offers great opportunity for the development and spread of disease, pests and parasites. The potential for survival in ecosystems is much enhanced in multi-species populations: the greater the species diversity in any assemblage, the better the chance will be of a balanced interrelationship between organisms. Man-created monocultures are thus ecologically unstable and can only be sustained at the price of high inputs of energy (e.g. machinery, weeding) or matter in the form of chemical fertilisers. Without these, loss of soil fertility quickly sets in, with all the dangers of soil erosion. Intensively managed monocultures can be very high-yielding, but only at the expense in the long run of drawing on energy supplies from non-renewable sources, the fossil fuels.

Eutrophication

Ecosystems cannot operate without efficient nutrient cycling. A major consequence of man's simplification of ecosystems is that he inevitably destroys major nutrient reservoirs, notably the natural vegetation and the soil system. To maintain yields he attempts to replace the loss by injecting fertilisers into the system.

When chemical fertilisers are applied to the land, many of the elements contained in them are retained by the soil, adding to the clay-humus complex. However, certain ions are not retained, and among them is nitrate, an important constituent of most fertilisers. Nitrate is being added to the soil from fertilisers and nitrogen-fixing plants at a much faster rate than it can be broken down by denitrifying agents in the soil. Being soluble, it is rapidly leached out into rivers and lakes. Here, the increased nitrogen input permits the accelerated growth of plants, algae and other phytoplankton: this chemical enrichment resulting in increased productivity is called **eutrophication**. Unfortunately, in extreme form the outcome is ultimately harmful, since the plants and organisms die and decompose at such a rapid rate that oxygen levels fall until aquatic life becomes impossible. A severe example of eutrophication has occurred in recent years in Lake Erie, North America, where deep layers of decaying organic matter have covered large stretches of the shoreline.

This example of man's impact on nutrient cycling in ecosystems is by no means unique: similar problems of eutrophication arise with the phosphates contained in detergents, fertilisers, and normal sewage effluent.

Effect on Individual Species

The extinction or reduction in numbers of plant and animal populations is a well-known consequence of man's impact on the environment. Often the species become endangered not so much by hunting or conscious elimination, but by the disruption and fragmentation of habitats. Some species, particularly large predators, require an extensive area of specialised habitat in which to breed and hunt, and fragmentation of this by man's interference has frequently had disastrous effects. The marsh harrier (*Circus aeruginosus*), a large raptorial bird of reedbeds and fens, is a prime example of this. The systematic draining and reclamation of fens and wetlands in Europe and North America has had a devastating impact on the population levels of the bird. Very few pairs have succeeded in raising fledgelings in Britain in recent years, and it seems likely that it will soon be lost as breeding species from the British Isles.

A contrary but equally far-reaching effect has been the accidental or purposeful introduction of alien species into ecosystems. Some animals and plants, because of their greater genetic adaptability and high reproductive rates have often made places for themselves at the expense of native species. Others have taken advantage of the new artificial ecosystems that man has created. For example, the starling (*Sturmus vulgaris*) and the rock dove (*Columba livia*) were once cliff-dwellers, but they have now established themselves in cities throughout the world, roosting in the 'cliffs' of city centres. The gradual extension in the distribution of the starling in North America (Fig. 24.4), following its introduction in New York in 1891, represents an interesting example of an artificially introduced species that has successfully competed with native birds for living space in a man-made environment.

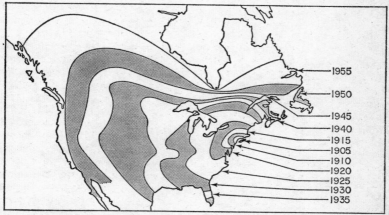

Fig. 24.4. Gradual extension of the distribution of the European starling in North America from 1905 to 1955.
(Reproduced from Cox, Healey and Moore:
*Biogeography: An Ecological and
Evolutionary Approach*, 2nd edition, 1976,
by permission of Blackwell Scientific Publications)

The Present Status of Ecosystems

We have seen that the evolution of all ecosystems is primarily determined by the amounts of energy and matter which flow through them in chains or webs. They are further maintained by intricate patterns of chemical cycling. Under natural conditions, ecosystems have been in a state of ecological equilibrium. With the increasing impact of man, their essential characteristics are altered, so that now signs of severe imbalance or a declining efficiency are beginning to be observed in many of them. This is shown, for example, by the progressive devastation of formerly good fertile agricultural or grazing land through over-intensive use; in the reduction of species when secondary forest replaces primary forest; in a general loss of biological productivity; and in an increasing amount of pollution.

Suggested Further Reading

Billings, W. D., *Plants and the Ecosystem* (Chapter 8), Macmillan, London, 1964.

Boughey, A. S., *Man and the Environment*, Macmillan, New York, 1971.

Jones, D. K. C., 'Man-made landforms', in *The Unquiet Landscape*, edited by Brunsden, D., and Doornkamp, J. C., IPC Magazines, London, 1974.

Smith, R. L., *The Ecology of Man*, Harper and Row, London, 1972.

Strahler, A. N., and Strahler, A. H., *Environmental GeoScience*, Hamilton, California, 1973.

Turk, A., *et al.*, *Environmental Science*, Saunders, London, 1970.

Wagner, R. H., *Environment and Man* (2nd edn.), Norton, New York, 1974.

CHAPTER TWENTY-FIVE

APPLIED PHYSICAL GEOGRAPHY

Much of man's impact on the environment is detrimental both to natural processes and systems and, in the long run, also to himself. The purpose of this chapter is to consider some of the more positive aspects of the situation: to see how physical geography can be made problem-orientated and employed to alleviate some of these problems as part of the need to conserve the natural environment for the benefit of all organisms, including man. The aspects we shall consider in this chapter are by no means comprehensive; the aim here is to indicate something of the range of application rather than to cover all the possibilities.

Although many environmental problems are clearly of man's own making, others arise from purely natural phenomena. Severe catastrophes such as floods or violent storms are generally referred to as **environmental hazards**. A considerable element of applied physical geography is concerned with dealing with these in an attempt to make living conditions safer. The physical geographer, with his concern for many aspects of the natural environment, is also in a unique position to contribute much to the field of **environmental management**, an integrated planning concept which seeks to rationalise man's demands on particular areas with the need to conserve the environment. Other applications of physical geography have **economic** overtones: weather variations, for example, have a strong bearing on the viability of many of man's economic activities. Finally, in this chapter, some of the problems associated with **natural resources** are considered: the best use of resources must rest on a thorough understanding of the total landscape, which physical geography attempts to provide.

Environmental Hazards

Man has always been subject to natural disasters over which he has little control. Most people would include volcanic eruptions, earthquakes, landslides, tornadoes, hurricanes and cyclones, storm surges, floods, droughts, blizzards and forest fires in a list of environmental hazards. It is worth emphasising that many of these only become hazards because man elects to use areas susceptible to these natural phenomena; this applies especially to earthquakes and floods. Clearly, what is far more difficult to assess is the seriousness of these hazards in any particular place, for much depends on people's **perception** of the hazard. Perception varies with culture and with time. For example, the possibility of cyclone damage has always existed around the Timor Sea, but one can say that the perception of the problem by the inhabitants of Darwin, northern Australia, changed following the disaster of Christmas Day, 1974. Before, buildings were mainly of light construction to facilitate air circulation: most of these were destroyed. Now, with the disaster still fresh in mind, the cyclone hazard is perceived in a different light and the city is being built on far more substantial lines. A further discussion of the problem of perception in geography may be found in Chapter Two of *Economic and Social Geography Made Simple*.

Depending on how the severity of the hazard is perceived, man has the choice of either ignoring it and accepting the risk of loss, or of doing something about it. In the latter case, studies within the scope of physical geography clearly have a role to play in providing information about the magnitude and frequency of these events, where they are likely to occur, and the processes causing them. An understanding of the forces involved gives man a chance of exerting at least some limited control over these hazards. As illustrations of what can be done, we can consider the examples of weather hazards and flooding.

Weather Hazards

In dealing with weather hazards, meteorological information is usually employed to predict their occurrence rather than to prevent them. Good weather observation networks are obviously of great importance here. Long-term meteorological records help us to identify the times and places of severe weather events, and to relate them to the occurrence of other climatic phenomena. For example, there seems to be a close relationship between hurricane and cyclone seasons and the position of the intertropical convergence zone (Chapter Thirteen). In the short term, weather observations from ground stations, satellites and aircraft can be used to track weather systems and give advance warning of their movement. This especially applies to violent storms. In the Caribbean a close network of stations exists to alert the Gulf States to likely hurricane hazards. Advance warning allows a number of measures to be put into operation, such as the evacuation of people and livestock from areas likely to be flooded. Avoidance of the hazard is often cheaper than the almost impossible task of making everything hurricane-proof.

The alternative way of responding to weather hazards, that of modifying the atmospheric processes, is still very much in the experimental stage. Most attempts have centred around the principle of **cloud seeding**, the supply of hygroscopic nuclei to accelerate the Bergeron precipitation mechanism (Chapter Eleven). Early experiments in the 1940s were largely concerned to induce rainfall by using dry ice particles or silver iodide smoke. More recently, severe drought in Florida in 1970 and 1971 led to an intensification of rain-making using a method of dropping pyrotechnic flares into the tops of cumulus clouds. A sevenfold increase in the rain to be expected from these clouds has been claimed. As a method of reducing drought hazards, it is limited because the technique is only successful where large unstable masses of moisture-laden air are already present.

Seeding has also been used to reduce the severity of hurricanes, relying on the theory that seeding causes the rapid condensation of supercooled liquid particles and drains off reserves of latent heat in the hurricane system. Other attempts have been made to modify hailstorms, which cause many millions of dollars worth of damage in the United States. Again, the principle is that of seeding: supplying vast numbers of nuclei for condensation can induce the formation of many small hailstones, rather than the fewer large stones which cause so much damage. Cold fogs have also been dispersed by seeding: since these consist of supercooled droplets that can be cleared using liquid propane or dry ice, the seeding causes rapid transformation into ice particles which fall to the ground.

Flooding

Flooding is one of the most ubiquitous of natural hazards to affect man, mainly because he chooses to live in so many flood-prone areas. The reasons for widespread flood-plain occupancy are many and varied: availability of flat land, access to water, lack of perception of the flood hazard, are but a few of these. In contrast to the response to hurricanes and cyclones, man is in a better position to prevent and control floods, since surface water is more amenable to interference than atmospheric processes.

Many aspects of physical geography dealt with in this book come into play when dealing with flooding as an applied problem, including climatic, geomorphological, hydrological and biotic factors. The close relationship within a drainage basin between systems operating on slopes and river response, provides an important theoretical base (Chapters Four and Five). In practice, there are two fundamental approaches to flood control. The first is to prevent the flood forming in the first place, or at least to diminish the amount of water reaching the rivers. This can be achieved by modifying the slopes in the upper catchment or by building small dams on fingertip tributaries. We noted in the previous chapter that some of man's activities on the land appear to have aggravated floods. Reversing these actions—for example, by reafforesting slopes to increase the amount of interception and infiltration —can have a large effect on run-off. The other approach is to control the flood in the channel. Normal practice is to build large dams, usually as part of multi-purpose schemes, or to modify the channel and its banks. Engineering measures such as retaining walls and river diversions can be introduced where they are needed, close to centres of population or valuable land. These measures in themselves can of course have important consequences on river velocity and the patterns of erosion and deposition.

In the United States, considerable arguments have developed about the relative effectiveness of these two approaches, and this has become known as the 'upstream–downstream controversy'. Ideally, flood control and administration should be geared to total catchment areas, but frequently several States may be involved, each reluctant to pay for measures not within their administrative boundaries.

Environmental Management

The term 'environmental management' is a very popular one today. It implies the sensible and sustained application of the principles of science and technology to the landscape, and so brings together a wide range of individuals, both environmental scientists and decision-makers. Geographers in theory ought to be in a good position to act in the role of both, or at least to take an overview. Here is it inappropriate to discuss the problems of decision-making (see *Economic and Social Geography Made Simple*). Instead, we can take a brief look at the contribution of the physical aspects of geography: this contribution can be very much wider than simply modifying environmental hazards. Physical geographers ought to be able to evaluate the nature of the interactions between the various systems in the natural environment and deal with individual problems in integrated management context. **Drainage basin** or **watershed management** is an example of this: it emphasises that river catchments are a natural unit for the operation of many physical and biological systems, including human activity. Although drainage basins do not exert a restrictive influence on man, it can be argued that they form a much better framework for environmental management than the administrative boundaries we are familiar with.

Landforms and Environmental Management

Earlier in this book we saw that geomorphology embraced the study of form, materials and processes acting at the Earth's surface. All three aspects play an important role in attempts to manage the land in an intelligent way. The awareness of the importance of geomorphology in environmental management is now growing rapidly after a very slow start. Broadly, there are two lines on which applied landforms studies are developing: first, **form and process studies** are being used in the solution of specific site problems; and second, **landform evaluation** is being used to contribute to resource surveys.

The value of understanding the dynamics of landform processes has been amply demonstrated many times in schemes that involve attention to erosion and deposition, be it on slopes, in rivers or at the shoreline. For example, any attempt to stabilise slopes must rest on a fundamental knowledge of the causes of instability and the types of mass movement (Chapter Four). At Folkestone Warren, Kent, investigations in the 1940s showed that the large Chalk landslips were primarily caused by erosion of underlying Gault clay by the sea, and that the slips were rotational in character. On the basis of this knowledge, the remedial action applied was aimed to prevent marine erosion on the one hand, and to stabilise the rotational tendency by weighting the foot of the unstable area with a large concrete apron.

Equally, examples can readily be quoted of improvement schemes where geomorphological principles were not applied or not fully understood. In the valley of the River Aare, Switzerland, attempts were made in the nineteenth century to extend the amounts of usable land on the valley floor. Engineering works were carried out on the river banks to alleviate flooding. However, in time, without recharge from floodwater, the reclaimed lands dried out, subsided and caused renewed flooding. Now a second phase of engineering work has been carried out, this time taking the dynamic aspects of fluvial processes much more into account.

Landform evaluation as an applied science had its beginnings in Australia in the 1950s. This early work largely involved the compilation of an inventory about various aspects of the landscape in which a landform classification provided the framework for storing knowledge about the physical landscape. This became known as the **land systems** approach. At the same time, in Poland attention was being directed towards a more comprehensive scheme for mapping not only form, but also materials and processes as a basis for planning. This technique of **applied geomorphological mapping** is a considerable extension of morphological mapping (Chapter Four), which is concerned with form alone. Geomorphological resource maps of the Polish type provide information of considerable value to land-use planning, hydrological engineering, soil survey and conservation, and in providing the context for specific geomorphological problems. They have not so far been used to any extent in Britain or North America.

Ecosystem Management

The widespread current concern over the status of ecosystems is the product of a movement that has slowly been gathering momentum over the last hundred years or so. Early efforts at the conscious management of biological resources often arose out of economic necessity: the imminent disappearance of the last forests in Britain and central Europe in the eighteenth century prompted landowners to adopt methods designed to save the last remaining forests and to stimulate forest productivity. The **conservation movement** as such was born in the United States. The ruthless clear-felling of great stands of forest in the western states and the experience of the Kansas dustbowl in the 1930s served to focus attention on the problem. Currently, of course, conservation has become a much wider issue than the maintenance of natural biological systems.

As applied to organic resources, one aspect of environmental management is the preservation and protection of wild life or of natural habitats from modification and depletion by man. This may be carried out for a combination of ethical, scientific or aesthetic reasons. To this end, nature reserves, wild-life refuges and similar controlled areas have been set up all over the world, designed to protect a particular habitat and its communities. These have not always been entirely successful. A classic example of the lack of understanding of ecological principles occurred with the establishment of National Parks in East Africa: these were designed originally to protect game animals, man being excluded except as a sightseer. But as a result, animals such as elephant, hippo and buffalo, whose populations had formerly been kept in check by hunting, increased to an extent that widespread devastation of their habitat resulted. What has been often overlooked in the past in environmental management is, first, that ecosystems cannot simply be 'preserved', but are dynamic in character, and second, man is an important habitat factor in many cases: the ecological niche occupied by him cannot suddenly be left vacant.

A second element of ecosystem management, one which has come much more to the fore in recent years, is that of maintaining sustained yield from organic resources. This idea was first applied to the maintenance of the breeding stocks of marine animals and to forestry practice. It is also implicit in the principles of **soil conservation**, the aim of which is to sustain agricultural fertility. Many authorities would maintain that this is by far the most important aspect of ecosystem maintenance, and that in the face of growing pressure on food resources, the protection of wild life for non-productive reasons is a luxury we can ill afford.

Modern environmental management policies attempt to reconcile these apparently conflicting aims—namely, preservation and productivity. Multi-purpose schemes are often now attempted. This is most successfully applied to management of forests, which because of their size are often well suited to a variety of uses—for timber, wild-life conservation, water supply and recreation. In Britain, National Nature Reserves are now managed as multiple resource units.

In summary, there clearly is a need to ensure that environmental management permits the maximum use of biological resources consistent with the maintenance of the greatest diversity of organic life.

Economic Value

Physical geography is sometimes seen as being of scientific interest, but of not much economic value. Obviously, it is possible to put some approximate value on individual resources, such as sand and gravel or a prized species of tree or animal. However, it is hoped that it is also clear from what has already been said in this chapter that the application of the subject has very considerable economic benefits in the long run. The prevention of natural hazards and the sensible management of the environment may need considerable short-term expenditure, but this has to be weighed against their long-term value, not only in cost-benefit terms, but for the quality of life.

In the case of plant and animal ecology, what is being increasingly realised is that the preservation of organic diversity is not only ecologically preferable, but economically desirable and profitable: an individual species in a mixed community may give a smaller short-term economic return than if cultivated in a simple monoculture, but this is compensated in the long run by greater stability and sustained yield using far less fertiliser and fossil fuel energy.

In the case of geomorphology, advanced knowledge of superficial deposits, ground-form and erosional and depositional processes at any particular engineering site could save constructors and taxpayers a lot of money. Even this short-term saving is far outweighed by the benefits of applying landform studies to check further flood damage, soil erosion and landslides, with all the social and economic loss they often entail.

The Value of the Weather

In contrast to geomorphology and ecology, weather study offers many more examples of tangible economic return, mainly because the benefits of application are more immediate. Although man is capable to a certain extent of creating his own micro-climate by building homes and heating them, weather and climate affect the economics of many of man's activities. Primary economic activities, such as farming and fishing, are particularly weather sensitive. Profits and loss in agriculture are often closely associated with the frequency of various elements—rain, hail or the number of sunshine hours—depending on the particular crop grown. Adverse weather conditions clearly have a monetary impact on fisheries in preventing catches being made; in addition, sea and air temperatures have an effect on the total fish population. Equally, in the tertiary sector of the economy, transportation costs are strongly influenced by weather variations. Fog may disrupt airline traffic, causing a substantial loss of revenue to airlines and airports; strong headwinds increase fuel costs. The construction industry is also liable to be affected by snow and heavy rain.

All these are fairly obvious examples. There are also a number of more subtle relationships between weather and economics. For example, a number of studies have demonstrated how temperature variations can affect the efficiency of workers in factories. Weather also affects the retail trade: the types of goods sold and the number of customers in department stores has been shown to vary with the type of weather. Finally, as many of us are acutely aware, our fuel bills for heating our homes depend to a large extent on the prevailing weather conditions.

There is demonstrably considerable economic value in weather study.

Specific investigations are called **econoclimatic** studies. These have been chiefly employed in relation to agriculture. No integrated study has yet been made of the monetary impact of weather variation on the total economy of a region.

Resource Evaluation

In the context of the twentieth-century concern over the rate of destruction and depletion of natural resources, we can conclude this chapter by looking at the environment from a resource viewpoint, and attempt to evaluate the consequences of continued use.

Like all other organisms, man is dependent on environmental resources for food energy and also for raw materials such as water and certain minerals. Some of the resources we need from the environment, such as solar energy, are virtually inexhaustible; others can be maintained or renewed with careful

Table 25.1. Exhaustibility and Renewability of Resources

Inexhaustible resources	Exhaustible but renewable resources	Exhaustible but irreplaceable resources

Total amounts of:
  Atmosphere
  Water
  Rock
  Solar energy
	Water in usable condition
	Vegetation
	Animal life
	Certain soil minerals
	Uncontaminated CO_2
	  and O_2 where needed
	Certain ecosystem types
		Soil
		Certain minerals
		Rare species
		Certain ecosystem types
		Landscape in natural condition
		Much of the ground-water
		  supply

Source: Costin, A. G. 'Replaceable and irreplaceable resources and land use', *Journal of the Australian Institute of Agricultural Science*, Vol. 25, pp. 3–9, 1959.

management. A classification of resources based on their degree of exhaustibility is suggested in Table 25.1. Strictly speaking, the total amounts of air, water and rock on earth are finite, but under existing conditions, the quantities are so great that there will always be enough to meet our needs.

Although there is a great deal of water on this planet, much of it is saline, and fresh or usable water is one example of a resource that is exhaustible, particularly in local situations. However, it can be maintained near where it is needed by proper management of the catchment area, including the building of dams and reservoirs. Another example of a potentially exhaustible resource is plant and animal life itself. In his search for continuing high yields from biological resources it is very easy for man to overexploit. Bitter experience has taught that overexploitation is likely to be irreversible in the tropics and arctic areas, but less dramatic in temperate zones. Because the reproductive capacity of most plants and animals is relatively high, there can be a certain amount of recovery of most ecosystems if the damage is not too severe or extensive in area. To give one example, a small area devastated by a forest fire will be gradually revegetated because undestroyed seed sources are near by. A large burned area will revegetate very slowly: near-by seed sources will have been destroyed, migration over any considerable distance takes time, and the area is liable to be exposed to severe environmental conditions since there will be a lack of vegetation to protect against wind, drying out and erosion.

Fresh, unpolluted air must also be regarded as an exhaustible but renewable resource. At the local scale, in certain meteorological conditions unpolluted air is quickly exhausted; the only way to conserve it is to eliminate the sources of contamination. Fortunately, on a global scale, fresh air is still abundant and atmospheric circulation replenishes the system.

Some essential parts of the environment are irreplaceable if lost. This applies to soils, certain minerals and the fossil fuels, and to rare species of plant and animal. Of particular concern is the loss of soil, for soil is the immediate source of essential mineral elements and water for plant growth: it has already been stressed that without soil there will be no food. Soil can take thousands of years to form, yet be destroyed in a matter of months by unchecked soil erosion. Of all the necessary environmental resources, soil is the one we can least afford to lose, and **soil conservation** therefore becomes of prime importance.

As far as mineral and fuel resources are concerned, here is a situation in which matter has been concentrated by geological process. When these resources are used by man, the reverse occurs, and the matter is transformed into a dispersed state. For example, the combustion of coal disperses an extremely dense concentration of hydrocarbons into the atmosphere as heat. This is lost to space as long-wave radiation and there is no way in which it can be recovered. With other non-energy minerals, such as lead, a certain amount of recycling can be achieved, as with the recovery of the lead contained in car batteries, thus prolonging the availability of the resource.

The other side to the resources problem is that of **world population**. The world population is increasing at a rate of something like two per cent per year, meaning that it will double within the next thirty-five years. Up to the twentieth century, the continued discovery and opening up of new lands to some extent kept pace with population increases in the western world, since fresh resources could be injected into areas or countries already depleted. Where fresh injection of material and energy has ceased, the quality of life has inevitably fallen. The fact that the standard of living is so low in many parts of the world is partly due to the exhaustion of such irreplaceable resources as the soil and a balanced natural landscape. However, it is also a result of insufficient or improper use of resources still there. Given proper management of the world's resources, we can take the optimistic view that the world's present population could be adequately fed. In the longer term, unless there is some major technological revolution, the limited carrying capacity of the earth is likely to prove increasingly inadequate.

Suggested Further Reading

Chorley, R. J. (ed.), *Water, Earth and Man*, Methuen, London, 1969.

Cooke, R. V., and Doornkamp, J. C., *Geomorphology in Environmental Management*, Clarendon Press, Oxford, 1974.

Ehrlich, P., and Ehrlich, M., *Population, Resources, Environment*, Freeman, San Francisco, 1970.

Maunder, W. J., *The Value of the Weather*, Methuen, London, 1970.

Tivy, J., *Biogeography* (Chapters 15 and 16), Oliver & Boyd, Edinburgh, 1971.

White, G. F. (ed.), *Natural Hazards*, Oxford University Press, London, 1974.

THE GEOLOGICAL TIME-SCALE

Era	Period		Epoch	Date of commencement (millions of years ago)	Major mountain-building episodes in Europe
CAINOZOIC	QUATERNARY		HOLOCENE	10,000 years ago	Alpine
			PLEISTOCENE	2	
	TERTIARY	NEO-GENE	PLIOCENE	7	
			MIOCENE	26	
		PALAEO-GENE	OLIGOCENE	38	
			EOCENE	55	
			PALAEOCENE	65	
MESOZOIC	CRETACEOUS			135	
	JURASSIC			190	
	TRIASSIC			225	
PALAEOZOIC	PERMIAN			290	Hercynian
	CARBONIFEROUS			340	
	DEVONIAN			400	Caledonian
	SILURIAN			430	
	ORDOVICIAN			500	
	CAMBRIAN			580	
PRECAMBRIAN			Oldest rock	3,500	
			Origin of Earth	4,500	

INDEX

INDEX